# AS/A-LEVEL

## STUDENT GUIDE

## WJEC/Eduqas

# Geography

Coastal landscapes

Tectonic hazards

Kevin Davis
Sue Warn

**HODDER**
EDUCATION
AN HACHETTE UK COMPANY

Although every effort has been made to ensure that website addresses are correct at time of going to press, Hodder Education cannot be held responsible for the content of any website mentioned in this book. It is sometimes possible to find a relocated web page by typing in the address of the home page for a website in the URL window of your browser.

Hachette UK's policy is to use papers that are natural, renewable and recyclable products and made from wood grown in well-managed forests and other controlled sources. The logging and manufacturing processes are expected to conform to the environmental regulations of the country of origin.

Orders: please contact Hachette UK Distribution, Hely Hutchinson Centre, Milton Road, Didcot, Oxfordshire, OX11 7HH. Telephone: (44) 01235 827827. Email: education@hachette.co.uk. Lines are open from 9 a.m. to 5 p.m., Monday to Friday. You can also order through our website: www.hoddereducation.co.uk.

© Kevin Davis and Sue Warn 2020

ISBN 978-1-5104-7215-0

First printed 2020

First published in 2020 by
Hodder Education,
An Hachette UK Company
Carmelite House
50 Victoria Embankment
London EC4Y 0DZ

www.hoddereducation.co.uk

Impression number 10 9 8 7 6 5 4

Year        2024

Cover photo: Mihai Andritoiu/Adobe Stock

Typeset by Integra Software Services Pvt. Ltd, Pondicherry, India

Printed and bound by CPI Group (UK) Ltd, Croydon, CR0 4YY

A catalogue record for this title is available from the British Library.

# Contents

## Content Guidance

### Coastal landscapes

### Tectonic hazards

## Questions & Answers

# ■ Getting the most from this book

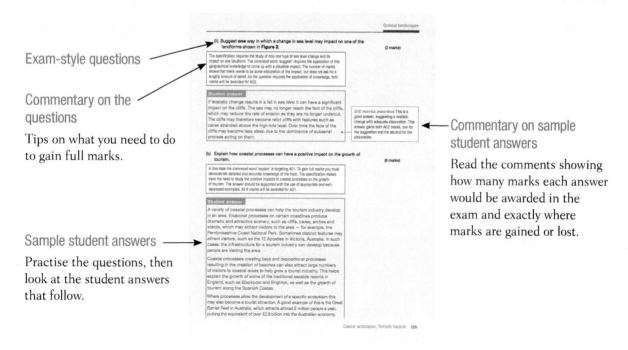

Exam-style questions

Commentary on the questions

Tips on what you need to do to gain full marks.

Commentary on sample student answers

Read the comments showing how many marks each answer would be awarded in the exam and exactly where marks are gained or lost.

Sample student answers

Practise the questions, then look at the student answers that follow.

# About this book

This guide has been designed to help you succeed in WJEC AS and A-level and Eduqas A-level Geography: **Coastal landscapes** and **Tectonic hazards**. The guide has two sections:

The **Content Guidance** summarises the key information that you need to know to be able to answer the examination questions with accuracy and depth. In particular, the meanings of key terms are made clear. You will also benefit by testing your knowledge with knowledge check questions, and noting the exam tips, which provide further help in determining how to learn key aspects of the course.

The **Questions & Answers** section includes sample questions similar in style to those you might expect in the exam. There are sample student responses to these questions as well as detailed commentary giving further guidance in relation to what exam markers are looking for in order to award top marks. The best way to use this book is to read through the relevant topic area first before practising the questions. Refer to the answers and comments only after you have attempted the questions.

The topics covered in this guide are:

**Eduqas A-level Component 1: Changing landscapes and changing places**
- Section A: Changing landscapes: Coastal landscapes

**Eduqas A-level Component 3: Contemporary themes in geography**
- Section A: Tectonic hazards

**WJEC AS Unit 1: Changing landscapes**
- Section A: Changing landscapes: Coastal landscapes
- Section B: Tectonic hazards

**WJEC A2 Unit 4: Contemporary themes in geography**
- Section A: Tectonic hazards

The formats of the different examination papers are summarised in the table below.

| Specification and paper number | Total time for Coastal landscapes/ Tectonic hazards | Total marks for Coastal landscapes/ Tectonic hazards | Structured questions | Extended response/ essay |
|---|---|---|---|---|
| Eduqas A-level Component 1: Section A Changing landscapes: Coastal landscapes | **50 min** in a paper lasting 1h 45 min | 41/82 | **Two** compulsory structured data-response questions Marked out of 13 | **One** question from a choice of two Marked out of 15 |
| Eduqas A-level Component 3: Section A Tectonic hazards | **40 min** in a paper lasting 2h 15 min | 38/128 | None | **One** question from a choice of two Marked out of 38 |
| WJEC AS Unit 1: Section A Changing landscapes: Coastal landscapes | **40 min** in a paper lasting 2h | 32/96 | **Two** compulsory structured data response questions Marked out of 16 | None |
| WJEC AS Unit 1: Section B Tectonic hazards | **80 min** in a paper lasting 2h | 64/96 | **Three** compulsory structured data response questions Marked out of 22, 24 or 18 | None |
| WJEC A2 Unit 4: Section A Tectonic hazards | **38 min** in a paper lasting 2h | 20/64 | None | **One** question from a choice of two Marked out of 20 |

# Content Guidance

## Coastal landscapes

# █ The operation of the coast as a system

The coastal zone is a dynamic open system. It has inputs and outputs of energy and materials, such as sediment. Every stretch of coastline has stores of materials and energy, and a wide range of processes operate as flows to move these, such as river currents, waves, ocean currents and tides, and atmospheric processes such as wind. These all contribute to dynamic change.

When using systems theory it is often difficult to define the boundaries of the coastal systems (how far inland, how far out to sea). Figure 1 shows the widely accepted **spatial boundaries** of the coastal system. Within the coastal system there are a number of subsystems, such as cliff and beach systems.

**Exam tip**

Always use appropriate geographical terminology and use a geographical dictionary to define specialist terms.

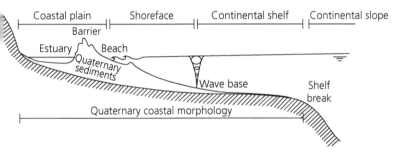

**Figure 1** Spatial boundaries of the coastal system

The most useful approach to studying coastal systems includes the **process–response method**, which states that the morphology of any coastal landform is a product of the processes operating in the system (these processes are driven by energy and sediments). **Coastal cliff retreat** is a good example to explain this approach (see page 23).

**Knowledge check 1**

What are the main inputs in a coastal system?

## The coastal sediment budget

Sediment and its movement are critical to the stability of a coastline. In order to manage a length of coastline it is important to know how much sediment is available, where it comes from, where it is stored, and how it leaves a particular coastal section. The identification of these factors is referred to as a **sediment budget** (Figure 2).

**Knowledge check 2**

Why can the coast be classified as an open system?

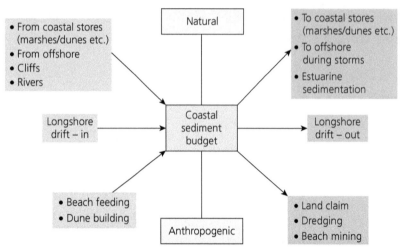

**Figure 2** The coastal sediment budget

It is difficult to estimate sediment inputs and outputs, especially when considering sediment movements from and to offshore stores. The significance of each source (input) and output varies according to different coastlines — for example, in 'soft rock' coastlines such as Barton on Sea (pages 55–56), cliffs are the major terrestrial supplier of sediment, but in other parts of the world, such as Bangladesh, fluvial sediment is the dominant supply.

Outputs to the coastal system include longshore drift, loss to offshore, and transfer to sediment stores down the coast. Large volumes of sediment can be temporarily lost to the sediment budget in stores and, similarly, aeolian transfer can be a short-term loss. As Figure 2 shows, anthropogenic losses from beach mining, dredging etc. can also impact on total sediment losses from the system (page 54).

In a balanced budget, input and output volumes should be in equilibrium:

volume of sediment in = volume of sediment stored + volume of sediment out

Human actions, such as building dams or hard engineering coastal defences, can upset the sediment balance, as the inputs suddenly decline. If a replacement sediment source cannot be found, then the following situation will occur:

volume of sediment in < volume of sediment stored + volume of sediment out

This produces a net loss of sediment to the budget, which accelerates erosion. Clearly, beach feeding or nourishment can represent a major input to help balance the sediment budget.

## Coastal sediment cells

In many countries, **sediment** or **littoral cells** have been identified as units of coastal management, where the dominant processes influencing the sediment budget are generally uniform within a particular coastal stretch. For example, in Wales there are three main sediment and littoral cells, with boundaries formed from the promontories at St David's Head, Bardsey Sound and Great Orme.

**Knowledge check 3**

Explain how beach nourishment can help to balance the sediment budget.

The movement of sand and shingle in the nearshore zone by littoral drift (longshore drift — page 27) has been found to occur in discrete, functionally separate cells. There are 11 major cells for England and Wales (Figure 3), with smaller subcells identified.

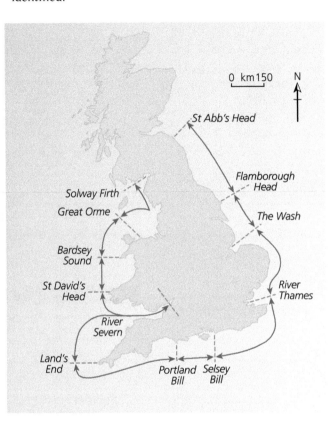

Figure 3 The coastal sediment cells in England and Wales

- A major cell is defined as 'a length of coastline and its associated nearshore area, within which the movement of coarse sediments is largely self-contained'.
- Sediment cells are functional systems because there is some movement across longshore drift divides. Littoral cells are therefore open systems.
- Sediment sinks (stores) occur where sediment transport paths meet.

## The concept of equilibrium

Coasts are dynamic zones of rapid change. These changes occur frequently and are largely caused by changes in energy conditions. For example, during storms the morphology of the coast responds to changes in energy because it aims to exist in a state of equilibrium, i.e. when the amount of energy entering the coastal system is equal to the energy dissipated. There are three types of equilibrium:

1 **Steady-state equilibrium:** where variations in energy and the morphological response do not deviate far from the long-term average. For example, where and when sea cliffs receive more or less equal

atmospheric and marine energy (e.g. from wind and waves), the profile of the cliff tends to stay the same from year to year, especially for resistant rocks. In the same way, a beach receiving similar amounts of wave energy from one year to the next undergoes seasonal adjustments, but its **average** annual gradient stays the same.

2  **Meta-stable equilibrium:** where an environment switches between two or more states of equilibrium, stimulated by some sort of trigger. For example, the actions of high-energy events, such as storms or a tsunami, which can remove a whole beach in hours, or human actions, such as the construction of a large breakwater or offshore dredging. This can rapidly switch a coastal system from one state of equilibrium to another, for example by removing or supplying large volumes of beach sediment.

3  **Dynamic equilibrium:** this also involves a change in equilibrium conditions but in a much more gradual manner than for meta-stable equilibrium, over a longer time period. A good example is the response of coasts to the gradual post-glacial eustatic rise in sea level, as large amounts of ice from ice sheets and ice caps have melted as a result of climate warming, so that wave energy actions occur higher up the shore, and cliff and beach profiles adjust as a consequence.

Equilibrium as a state does not apply to all coastal areas. Energy environments can change within just a few metres, spatially as well as temporally, which further complicates the issue.

## System feedbacks

Understanding states of equilibrium requires some knowledge of feedbacks within the system. Feedbacks occur as the result of change in a system, and they can be either positive or negative, switching the system to a new state of equilibrium or attempting to recover to the system's original state of equilibrium, respectively.

**Positive feedbacks** therefore **amplify** the initial change in the system so that, for example, the ridge of a coastal sand dune breached by storm wave erosion may be subsequently laterally undercut by wind erosion, thus fragmenting the dune ridge and leaving it susceptible to further wave erosion. Ultimately the whole dune ridge may be driven further inland and a new state of equilibrium reached.

**Negative feedbacks** diminish or dampen the effect of change. For example, sand eroded during a storm from the front of the embryo and fore dunes at the back of the beach may be redeposited offshore as sand bars, which then help to protect the beach dune system from erosion by slowing waves and dissipating the wave energy reaching the dune front.

Human intervention often leads to apparently unforeseen and undesirable feedbacks, often as a result of inappropriate coastal management.

**Knowledge check 4**

Explain how the coastal system is in a state of dynamic equilibrium.

**Exam tip**

There are certain key concepts that you must understand. These include systems, equilibrium and feedback.

**Exam tip**

The term negative feedback does not mean it has a detrimental effect on the coastal system. It is negative because it reduces the impact of the original change.

## Summary

- The coastal system includes inputs, outputs, stores and transfers of energy and materials.
- The sediment budget refers to the amount of sediment available, where it comes from, where it is stored, and how it leaves a particular coastal section.
- Sediment cells are units of coastal management where the dominant processes influencing the sediment budget are uniform within a stretch of coast.
- Coasts are dynamic zones of rapid and frequent change, largely caused by changes in energy conditions, for example during storms.
- Dynamic equilibrium involves a gradual change, over a long time period.

# ▋Temporal variations and their influence on coastal environments

Tides, currents and waves are key inputs of energy into the coastal system as they have the potential to erode, transport and deposit material. All three are rapid processes, in that they operate from instantaneous through to annual timescales but rarely on decadal scales, except in terms of cumulative change.

## Tides

**Tides** result from the gravitational attraction on water of the Moon and the Sun, with the Moon having twice the impact of the Sun. All coasts are influenced to some extent by tides, but only a few types of coastline, such as **lowland sandy estuarine coasts**, can be said to be tide-dominated.

## Tidal frequency

Most coastlines, such as all open Atlantic coastlines, experience **semi-diurnal** tides, i.e. two high and two low tides approximately every 24 hours. However, some places, such as Antarctica, receive genuinely **diurnal** tides — one high tide and one low tide each day, as a result of local factors.

As the respective motions of the Earth, Moon and Sun go through **regular cycles**, the gravitational forces change and, therefore, so do the tides.

Twice a month the Sun and Moon are aligned, so that their gravitational forces are combined and therefore there is a strong gravitational pull — this leads to above-average tides called **spring tides** that, if combined with strong winds and storms, can cause significant landform changes high up on the shore. Twice a month the Sun and Moon are at right angles with respect to the Earth. As their gravitational forces act in different directions, the overall effects are lessened, so lower-than-average tides result, called **neap tides**. On a biannual basis the largest of the spring tides occur at the vernal (spring) and autumnal equinoxes, i.e. when the Sun is 'overhead' at the Equator.

If you look at tide tables you will see that high and low water times in most parts of the UK are separated by more than six hours, so that high and low tides occur slightly later each day.

The overall effect of tides is seen in smaller systems because of variations in ocean depth, uneven seabed topography, with deep basins separated by shallower continental shelves and continents, and the shape of these landmasses. Because of the Earth's rotation (Coriolis force) tides circulate around **amphidromic points**. High and low tides therefore occur at different times along coasts, such as in the UK, as water swirls around the amphidromic points.

## Tidal range

The height difference between high water and low water during the monthly tidal cycle is known as the **tidal range**. Annually, this is highest at spring tides and lowest at neap tides. At an amphidromic point the tidal range is zero and it increases with distance from the point, so that the maximum range is along coastlines far from the point.

Tidal range is important for coastal geomorphology because it influences a number of physical processes:

- Tidal range determines the vertical distances over which coastal processes operate, especially wave activity. On a micro-tidal coast (tidal range of less than 2 m, e.g. eastern Australia), wave breaking is concentrated in a narrow vertical zone throughout the tidal cycle, so well-defined erosional features, such as wave-cut notches, are formed at the foot of cliffs (page 23). On a macro-tidal coast (tidal range in excess of 6 m, e.g. most of the UK coast, and parts of North America, such as the Bay of Fundy, which has the greatest tidal range in the world), wave energy is distributed over a wide area, so its erosional capacity is relatively less, resulting in more depositional features.
- During the diurnal rise and fall of tides, wetting and drying of the substrata occurs. In a macro-tidal environment, more substrata are exposed or submerged and therefore affected by processes such as salt weathering. Sand dunes are more likely to develop in a macro-tidal environment, as here there is a wide expanse of deposited beach sand, which dries and can be subsequently transported landward by aeolian processes (page 31).

## Currents

Currents are clearly identifiable water flows operating in the coastal zone.

**Tidal currents** are associated with tidal movements. As water rises with the tides it produces tidal currents. It floods the intertidal zone (**flood tide**), causing entrainment and deposition of material. The falling tide is the **ebb tide**, which carries material in the reverse direction. These tidal currents have maximum velocity at their midpoints, while at high and low water, current velocity slackens considerably, leading to sediment deposition. As with rivers (Hjulström curve), there are critical current velocity **thresholds** for tidal currents to transport different particle sizes. Therefore,

**Knowledge check 5**

Explain the term 'threshold' in the context of sediment entrainment and deposition at the coast.

muds are usually found in low-energy low and high intertidal areas, while sand shoals occur in mid intertidal areas. Overall there is considerable transfer of sediment from offshore to nearshore and back again within the intertidal zone on a daily basis during every ebb–flood tidal cycle.

**Shore normal currents** occur where waves approach the shore with their crests parallel to the coastline. Water is carried up the beach but there has to be a return flow — **rip currents** occur in fairly evenly spaced locations along the beach — they flow back at speeds of up to $1\,\mathrm{m\,s^{-1}}$ through the advancing waves. Note the contrasts with oblique waves, with **longshore currents** being responsible for longshore drift (page 27). **River currents** are also powerful currents, transporting both river sediment and fresh water into the coastal zone. Energetic flows are periodic and associated with high levels of river flood discharge.

## Constructive and destructive waves

The classification of waves is based on their geomorphological action and the way in which they break upon the shore.

Landforms are the result of inputs of energy to surface and near-surface materials. At the coast, the main source of energy is waves generated by the wind. The frictional effect of the wind on the seawater surface produces motion in the upper layer of the water (Figure 4). This motion is only a wave shape on the water surface and not an actual forward movement of water. Sea waves are therefore caused directly by wind, with individual water particles moving in a circular path as the wave shape moves across the water surface. The stronger and more persistent the wind, the higher the amplitude of the waves and the more energy there is available for 'work'.

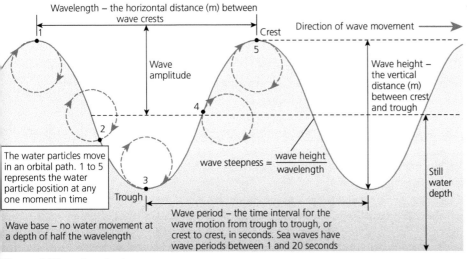

**Figure 4** Wave terminology

As a wave reaches shallower water at the coast, the circular water motion in the wave shape is affected by the sea floor. As water depth decreases (Figure 5) the water path movements change from a circular to an elliptical shape: wavelength and velocity decrease and wave height increases. The wave steepens and then breaks on the shore to produce different types of breaking wave (Table 1), depending on the ocean floor gradient.

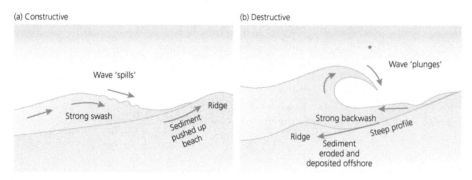

Figure 5 Constructive and destructive waves

Table 1 Features of constructive and destructive waves

| Feature | Constructive waves | Destructive waves |
|---|---|---|
| Wave height | Less than 1 metre | Greater than 1 metre |
| Wavelength | Long — up to 100 m between crests | Shorter — 20 m between crests |
| Wave period | Longer — six to eight breaking each minute | Shorter — 10 to 14 breaking each minute |
| Wave steepness | Gentle — tend to spill over Break at some distance from the shore Foam forms at the wave crest and becomes a line of surf as the wave approaches | Steep — tend to curl over and plunge down onto the shore |
| Wave energy | Low | High |
| Stronger swash or backwash? | Swash | Backwash |
| Movement of material | Up beach to form a berm | Down beach |
| Beach gradient | Increase gradient of upper beach and form a berm | Create a steep upper beach (storm beach) Reduce the gradient down beach |

Destructive waves are usually more frequent in winter as a result of the stronger winds associated with stormier weather. In the UK this is due to more frequent and intense Atlantic depressions, which are driven over the UK by the jet stream.

This variation in wave type, often on an **annual** basis, has an impact on sediment mobility and coastal morphology — the extent of wave energy dissipation determines whether coasts become erosional, depositional or stable.

**Exam tip**

Destructive waves do not necessarily result in erosional landforms. The term refers to the movement of sediment down a beach and its deposition offshore.

## Summary

- Tides, currents and waves input energy into the coastal system as they erode, transport and deposit material.
- All coasts are influenced by tides, which result from the gravitational attraction on water of the Moon and the Sun, with the Moon having twice the impact of the Sun.
- Most coastlines experience semi-diurnal tides, but some places have diurnal tides.
- Spring tides occur when the Sun and Moon are aligned, combining their gravitational forces to create a strong pull. Neap tides occur when the Sun and Moon are at right angles to the Earth, and their gravitational forces act in different directions.

- Currents are identifiable water flows, which include tidal currents, shore normal currents, onshore and offshore currents, longshore currents, rip currents and river currents.
- Destructive waves are associated with stronger winds, often in winter, whereas constructive waves are associated with less powerful winds in summer.

# Landforms and landscape systems, their distinctive features and distribution

## High-energy rocky coastal environments

The morphology and behaviour of rocky coastlines reflects the interplay between geological factors (rock structure and lithology (rock type)), and weathering and erosional processes.

Rocky coasts are **erosive** coasts, and are continually being cut back. While there is a comparatively slow rate of change averaged over long periods, recent research has shown that rocky coasts can be dynamic over short timescales, such as during winter storms, so they are called high-energy coasts.

Rocky coast erosion is accomplished by a wide range of processes working together. These processes can be grouped into three main types: mass movement, rock breakdown processes and marine 'rock removal' processes.

- **Mass movements**, such as rockfalls or landslides, depending on the rock type, are common because of the prevailing steep, and therefore unstable, slopes. All types of mass movement are **episodic**, occurring more commonly in winter as a result of more powerful waves undercutting the cliff base.
- **Rock breakdown processes** are physical, chemical and biological processes (types of **weathering**) that weaken and loosen rock material, making it available for removal by marine processes.
- **Rock removal processes** depend on the action of waves — energetic wave conditions in a concentrated micro-tidal environment are particularly effective, both in causing abrasion and hydraulic action. The larger and more powerful the waves, the greater the efficiency of the cross-shore and longshore sediment transport processes, driven by tidal currents, in removing loose material and keeping rocky coasts 'fresh' from debris.

Examples of high-energy coastal landforms include cliffs (page 22) and wave-cut notches (page 23). Classic examples for the study of rocky coastlines can be found in Dyfed (Pembrokeshire) or southern Gower in Wales, as well as in many places around the world, such as the Great Coast Road in southern Victoria, Australia. They are frequently classified as wave-dominated coastlines.

> **Knowledge check 6**
>
> Explain what is meant by 'episodic events', such as landslides.

## Low-energy sandy coastal environments

Low-energy coastal environments tend to produce largely depositional coasts with constructive waves pushing sediment shorewards, building sandy and shingle beaches that are often backed by sand dunes, with accompanying depositional features such as spits, bars and tidal mudflats. Low-energy environments can be macro- or micro-tidal, but have shallow shoreline gradients.

While low-energy conditions are prevalent, they are frequently interspersed by dramatic storm events that upset the dynamic equilibrium and lead to major changes in beach and bar morphology. On a stretch of lowland coast you will find both wave-dominated coastal stretches of beaches, barriers and dunes, and tide-dominated stretches of coast, usually around estuaries.

Examples of low-energy coastal environments are widespread, often interspersed with rocky shorelines. Extensive stretches of lowland low-energy coastlines occur along the German Baltic coast, the Netherlands coast, much of southern and eastern USA and the Lagos coast of Nigeria.

### Summary

- High-energy rocky coastlines are erosive, continually being cut back by mass movement, rock breakdown and rock removal processes.
- Erosional landforms include cliffs, shore platforms and wave-cut notches.
- Low-energy coasts are largely depositional, with shallow shoreline gradients and constructive waves pushing sediment shorewards.
- Depositional landforms include sandy and shingle beaches, sand dunes, spits, bars and lagoons, tombolos and cuspate forelands.

# ■ Factors affecting coastal processes and landforms

## Waves

The essential features of a wave were defined on page 13 (see Figure 4). Most of the waves affecting coastal zones are entirely wind generated (except tsunamis and storm surges), caused by the frictional drag between wind and the water surface. The amount of energy transferred between the wind and the water depends on:

- wind velocity
- wind duration
- fetch (the distance over open water that the wind blows to generate the waves)
- orientation of the coast to the waves

The greatest energy transfer occurs when strong winds blow in the same direction over a long distance for a long period of time, for example the great swell waves driven across the Atlantic by the prevailing southwesterly winds. Certain coasts

### Knowledge check 7

Define the following aspects of wave terminology: wavelength, amplitude, frequency.

receive high-energy waves — this includes all of the west coast of the UK — as they are orientated towards these waves.

Waves bring huge amounts of energy into the coastal system. As they approach the land, they are modified as a result of the decreasing water depth. Different types of wave are described in Table 1 on page 14.

The shape of the coastline and its orientation to the oncoming waves can affect the impact of waves. Although a coastline may be at the receiving end of a long fetch, its orientation may protect it from the full impact of high-energy waves, with the waves arriving perpendicular to the coast having a different impact from those arriving at an oblique angle.

## Wave refraction

As waves approach a coastline, the direction of wave approach is modified by the topography of the seabed. Wave refraction causes the wave energy from the breaking waves to vary along the coastline (Figure 6). This wave energy becomes concentrated at headlands and disperses around bays. The complex interaction of coastal configuration, offshore topography, exposure, wave fetch and landforms helps to explain variations in processes and landforms. For example, waves are refracted around a spit to give it a curved end or hook (page 29). You can sketch wave refraction patterns for any chosen stretch of coastline using Google Earth images.

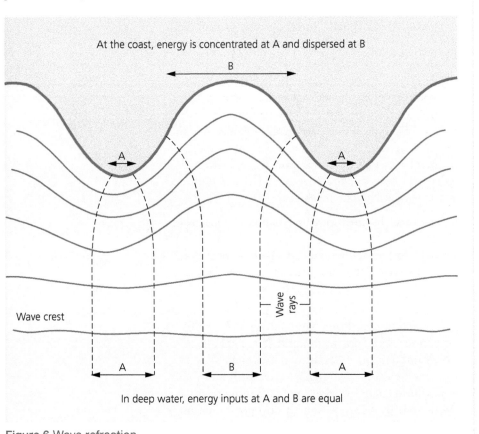

**Figure 6** Wave refraction

## Wave reflection

Along rocky coastlines, where there is deep water offshore, the waves are reflected back from the cliffs (known as the **clapotis** effect) and do not break at the shoreline. Wave reflection can also take place on very steep beaches or against sea walls. The interaction of reflected and incoming waves creates a **standing wave**.

**Knowledge check 8**

Explain how the clapotis effect affects cliff recession rates.

# The lithology of the coastline

A number of factors combine to make lithology (the make-up of the rock) an important influence on coastal processes and landforms.

## Hardness

As a result of heating and compression during their formation, igneous rocks (e.g. granite) and metamorphic rocks (e.g. gneiss) are harder and therefore more resistant to erosion. These types of rock form many high cliffs in northwest Britain. In contrast, many of the rocks that form the coastlines of southern and eastern Britain are 'soft rocks' — unconsolidated sand and clays of Tertiary age, as well as deposits of glacial boulder clay and gravels. Other factors being equal, these 'soft' rocks tend to be easily eroded, especially if the cliff bases are poorly protected by beaches, with an average recession rate of 3–6 m per year (after a storm, 10–25 m per year is not uncommon). An extreme example, with erosion rates of 30 m per year, has been recorded in the ash from the Krakatoa eruption in coastal Sumatra.

## Chemical composition

Chemical composition includes mineral composition and solubility. Some rocks, such as quartzite or most sandstones, are made almost completely from silica, which is chemically inert. The very low rate of chemical weathering adds resistance to the rock. Other rocks are more prone to rapid chemical weathering because of their chemical composition. Iron compounds oxidise in some sandstones, and feldspars are altered into clay minerals by **hydrolysis** in rocks such as granite. These 'rotted zones' increase vulnerability to both marine and subaerial processes.

The chemical decomposition of limestones by **carbonation** (solution) happens even more rapidly. It is caused by saltwater corrosion, leading to accelerated disintegration of some shore platforms. The impact of salt water not only affects limestones but also causes basalt to weather 14 times more rapidly than under freshwater conditions.

## Permeability

A further lithological factor is permeability, resulting from, for example, pores in an open-textured limestone or fissures and cracks or joints (e.g. in chalk or limestone). As surface water seeps through the cliffs, it increases the rock's resistance to subaerial processes, thereby adding strength to relatively soft rocks. This explains why chalk invariably forms high, near-vertical cliffs (such as at Beachy Head, Sussex, or the White Cliffs of Dover, Kent) and supports natural arches and stacks.

# The structure of the coastline

'Structure' can be defined as the way rocks are disposed or geologically arranged. Folding and faulting therefore provide a range of rock types with lithologies of different resistance to subaerial and marine processes, with less-resistant rocks being more easily exploited.

## Joints, faults and folds

The prevalence of joints, bedding planes and faults in a rock has a significant impact on the rates of weathering (both freeze–thaw and chemical). Where joints and bedding planes occur at high densities this weakens the rock and makes it subject to increased subaerial and marine erosion.

Faults or isolated master joints can be exploited by the sea to form a range of micro-features, as at Tintagel in North Cornwall (Figure 7). Narrow inlets (geos), such as those on the Island of Skomer, Pembrokeshire, develop along faults, which are zones of weakness (shatter zones). Folds (anticlines and synclines), where the rocks are stretched or compressed, also form weaker areas, for example at Saundersfoot near Tenby. Folding can also affect the angle of the bedding planes (dip), which can influence a cliff profile and the rate of erosion.

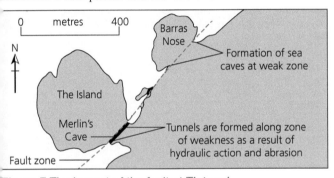

Figure 7 The impact of the fault at Tintagel

**Exam tip**

A case study of the Pembrokeshire coast is available at www.curriculum-press.co.uk.

In conclusion, the combined impacts of structure and lithology play a major role in determining both major and minor landforms, especially where coasts are composed of 'hard rocks'.

**Exam tip**

Be specific in your answers. Instead of naming lithology as a factor, state what aspects of the lithology have an influence, such as hardness, joints, folding or a combination of factors.

## Summary

- Coastal processes and landforms are affected by wave characteristics, and by the lithology and structural geology of the coast.
- Wave characteristics, including fetch, wave type, wave energy, orientation, refraction and reflection, all affect coastal processes. Wave refraction causes wave energy to be concentrated at headlands and dispersed around bays.
- The lithology of the rock influences differential erosion, transportation, deposition and coastal landforms.
- The rock structure — the effects of folding and faulting, joints, bedding planes and faults, and their arrangement — influences the shape and distribution of landforms.

# ■ Processes of coastal weathering, mass movement, erosion and the associated landforms

## Subaerial weathering

Although the same range of weathering processes occurs at the coast as inland, the presence of corrosive seawater and the daily effect of tides wetting and drying rock bring additional destructive influences. The key influences on the type and rate of weathering are geology and climate.

**Physical/mechanical weathering** breaks off rock fragments of varying sizes, which fall to the foot of the cliff where they can protect the cliff from erosion (Table 2).

Table 2 Effects of physical/mechanical weathering processes on coasts

| Weathering process | Effect |
| --- | --- |
| Saltwater crystal growth | Crystals (e.g. salt) grow when seawater that collects in cracks in the cliff face evaporates. As they grow, crystals exert pressure on the rock (most important) |
| Freeze–thaw | Repeated freezing and thawing of water causes a type of crystal growth that is most effective on high-latitude coasts with significant precipitation |
| Wetting and drying (water-layer weathering) | Expansion and contraction of minerals is most effective on clay and in macro-tidal environments |

**Chemical weathering** acts to decompose rock (certain types of rock are more susceptible) by altering the minerals in the rock (Table 3).

Table 3 Effects of chemical weathering processes on coasts

| Weathering process | Effect |
| --- | --- |
| Solution | Solubility of minerals depends on temperature and acidity of the water. Limestones are affected by carbonation, although they may be less soluble in seawater. Spray charged with carbonic acid leads to honeycomb weathering |
| Hydration | Minerals absorb water, weakening their crystal structure. Rock is then more susceptible to other weathering processes |
| Hydrolysis | Reaction between mineral and water related to hydrogen ion concentration in water, which particularly affects feldspar minerals in granite |
| Oxidation/reduction | Adding or removing oxygen. Oxidation results from oxygen dissolved in water and particularly affects rocks with a high iron content. Reduction is common under waterlogged conditions |
| Chelation | Organic acids, produced by plant roots and decaying organic matter, bind to metal ions, causing weathering |

Biotic weathering by plant roots is particularly active on vegetated upper slopes of cliffs, and opens up the cliff face to other destructive processes. Burrowing animals cause weathering of soft rocks such as clay and can also disturb coastal sand dunes.

# Mass movement

Mass movement can be defined as 'the downslope movement of material under the influence of gravity'. The type of mass movement that occurs is closely related to the geological structure and lithology of the coastline, the weathering processes that loosen the material on the cliff face, as well as the controls imposed by climate, past and present. Vegetation can also have a retarding impact.

Many instances of mass movement, which can occur very suddenly (within minutes), are triggered by undercutting by wave action at the base of a coastal cliff.

There are four types of *rapid* mass movement:

1 **Rockfalls**. Blocks of rock, dislodged by weathering, fall to the cliff foot. For example, in Svalbard (a high-latitude periglacial region) rock blocks are loosened by freeze–thaw action.
2 **Rockslides**. Blocks of rock slide down the cliff face, especially where rocks are dipping steeply towards the sea. This is common on the carboniferous limestone cliffs of Tenby and Gower in South Wales.
3 **Rock toppling**. Blocks or even columns of rock, weakened by weathering, fall seawards. This has occurred in the columnar basalt of the Giant's Causeway in County Antrim, in Northern Ireland.
4 **Rotational slides and slumps**. In a rotational slide, sections of the cliff give way along a well-defined concave slip surface. The fallen material stays as an identifiable mass off the shore because it is often composed of cohesive clays or boulder clay, so it may take a month for the sea to erode it.

**Slumps** occur where permeable rock overlays impermeable rock, such as at Christchurch and Barton on Sea (southern England). They are also common in unconsolidated rock (e.g. the sandy boulder clay deposits of north Norfolk and Holderness, both in eastern England). The subaerial processes are particularly active after periods of heavy rainfall, which lead to saturation and subsequent lubrication. A critical threshold is reached, triggering mass movement, slope failure and mudflows.

**Slow** mass movements include:
- **Creep** — the extremely slow (imperceptible) downslope movement of regolith (the loose material above the bedrock, including soil).
- **Solifluction** — the slow downslope movement of regolith, saturated by the melting of the active layer above the permafrost (in the summer of periglacial climate regions). There is some evidence that many parts of the UK experienced this type of movement in the last Ice Age.

**Exam tip**

Be clear about the differences between weathering and erosion, and between subaerial and marine erosion. Do not just use general terms such as erosion. Refer to the specific processes involved.

**Knowledge check 9**

Explain the difference between rotational slides and slumps.

# Marine erosion

A number of wave erosional processes exist, which work in combination.

- **Hydraulic action** results from waves breaking on bedded, jointed or faulted rocks, which can create hydraulic pressure in these structural voids. This can lead to weakening and readying of the rocks for further wave action.
- **Quarrying** occurs where powerful waves remove loose blocks.
- **Attrition** occurs when detached rocks break down through rubbing and banging against each other, gradually rounding the blocks as well as reducing their size.
- **Corrasion/abrasion** occurs when the material provided by attrition and abrasion is used by waves to further erode the rock.
- **Corrosion** occurs when rocks such as limestones are chemically attacked by waves.

# Characteristics and formation of coastal erosional landforms

## Cliffs and shore platforms

A cliff may be defined as 'any coastal slope affected by marine processes', although subaerial processes also play a significant role in cliff formation.

The cliff profile is determined by a number of factors. A variation in the balance of the inputs can result in different cliff morphologies in different locations (Figure 8). Cliff morphology can be determined by the following:

- **Geology**. The hardness and structure of the rock are very important in the formation of cliffs, with igneous and metamorphic rocks, and some sedimentary rocks such as limestone and sandstone, invariably forming steep cliffs. In contrast, unconsolidated rocks such as clays and sands usually result in low-angled cliffs. However, **constant** marine erosion at the cliff base can lead to frequent slope failure and, if the debris is removed, can again leave a steep slope ready to be undercut once more.

  Steep cliffs are also associated with either horizontal or vertical geological structure. The angle of inclination of the bedding planes (dip) is important, with seaward-dipping cliffs having low angles and landward-dipping cliffs having near-vertical faces.
- **The balance between marine erosion (wave activity) and subaerial processes**. Sufficiently energetic waves, usually highest along mid-latitude coasts, are required not only to erode the cliff but also to remove the debris created by wave erosion, as piles of subaerial debris tend to protect the cliff.
- **Inherited characteristics**. The sea may rework steep slopes initially formed by non-marine processes under different sea level situations. For example, some plunging cliffs that rise abruptly from deepwater fjords were originally sides of submerged glaciated valleys.

> **Exam tip**
>
> Make sure you are able to explain why the cliff profiles shown in Figure 8 differ.

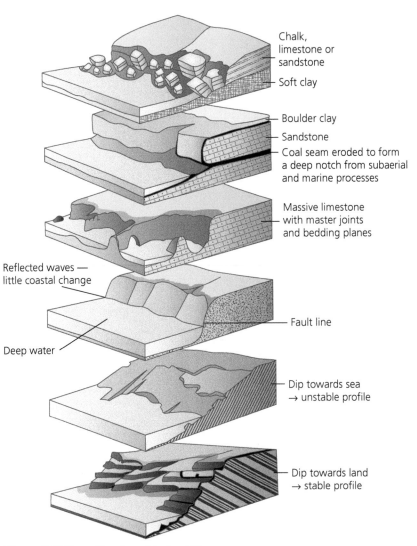

Chalk, limestone or sandstone

Soft clay

Boulder clay

Sandstone

Coal seam eroded to form a deep notch from subaerial and marine processes

Massive limestone with master joints and bedding planes

Reflected waves — little coastal change

Fault line

Deep water

Dip towards sea → unstable profile

Dip towards land → stable profile

**Figure 8** Cliff profiles found in the UK

## Cliff retreat and the formation of shore platforms

The sequence of cliff retreat (Figure 9) often proceeds as follows:

- A **wave-cut notch** is formed by wave quarrying and corrosion at the base of the cliff, which effectively undermines the cliff, causing slope failure either via slumping (soft rocks) or vertical cliff collapse.
- Where cliffs are fronted by a narrow shore, a cycle of notch formation, cliff failure, debris removal and cliff retreat takes place.
- In time, a wide shore platform develops. The cliffs are then no longer within reach of any marine action other than storm waves at the highest spring tides, so in time the cliff profile becomes degraded.

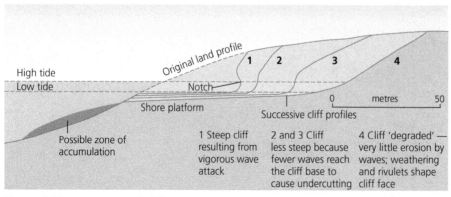

**Figure 9** The sequence of cliff retreat and the development of a shore platform

The combined impact of the causes of cliff recession (Figure 10) on cliff geology plays a key role in affecting rates of cliff retreat, which vary considerably, as shown in Table 4.

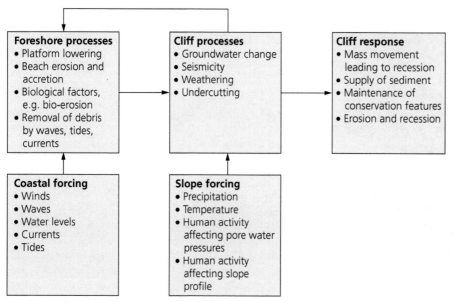

**Figure 10** The key causes of cliff recession

**Table 4** Average rates of cliff retreat by rock type

| Granites | Limestones | Chalk | Shales | Clays | Unconsolidated glacial deposits |
|---|---|---|---|---|---|
| Resistant | 1 cm per year | 10 cm per year | 1 m per year | 10 m per year | 100 m per year |

**Note:** this is a log scale

## Shore platforms

Intertidal shore platforms (sometimes known as wave-cut platforms) are created by wave quarrying and abrasion, but bio-erosion and salt weathering during tidal exposure are also significant — hence the abandonment of the term 'wave-cut platform'. Shore platforms are relatively flat expanses of gently sloping (usually 1–5° seaward) rock, found at the foot of a cliff and extending out to sea. Shore platforms also show the influence of rock structure (where the rocks are steeply dipping and

**Knowledge check 10**

Explain why the term 'wave-cut platform' is an inaccurate one and has been superseded by 'shore platform'.

esistant, the platforms are narrow and ridged) and lithology (differential erosion leads o variations of micro-relief), for example the ridges caused by resistant igneous dykes on the shore platforms of Arran in Scotland. Platform width is ultimately finite, as ncreasingly wide platforms dissipate wave energy before it reaches the cliff base.

## Headlands and bays

**Headland** and **bay formation** results from the differential erosion of juxtaposed rocks of varying resistance, especially where the coast is **discordant**, with the structural trend at approximately right angles to the coastal trend. This is in contrast to **concordant coasts**, where **cove** formation is the key feature. Clear examples can be found along the 'classic' stretch of coastline of Dorset's World Heritage Coast (Figure 11).

> **Exam tip**
>
> Differential erosion is a key concept. Use the case study of headland and bay formation along the World Heritage Coast of Dorset to identify name examples.

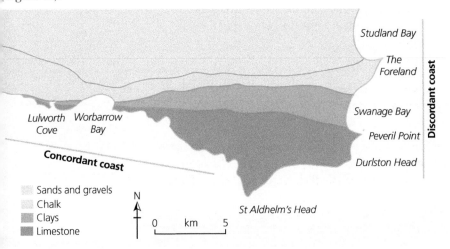

Figure 11 The role of geology in the formation of headlands, bays and coves on the Isle of Purbeck coast, part of the World Heritage Coast, Dorset

## The formation of micro-features

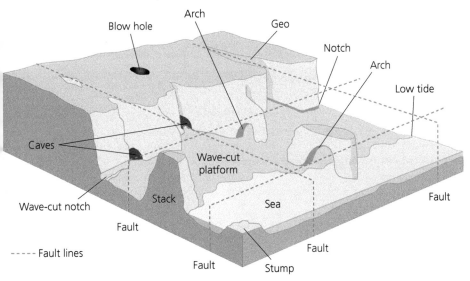

Figure 12 Cliff architecture

There are *two* likely sequences of development. Initially, **sea caves** develop (Figure 12), their distribution controlled by geological weaknesses, such as master joints, major bedding planes and faults.

## Sequence 1

The impact of air and water forced up into the caves by wave action can lead to the development of vertical shafts and tunnels upwards to the ground surface to form a **blow hole**, for example Spouting Horn in Kauai, Hawaii.

Air and water are forced through the blow hole by breaking waves, at certain tides and particular wind directions, with an explosive force. This causes large pressure changes in the cave and further erosion.

The blow hole roof may collapse to form a **geo** or inlet. Alternatively, differential erosion may exploit the weakness of the fault or shatter zone to form a very long, narrow gully, or geo, for example in Orkney.

## Sequence 2

Differential erosion may result in adjacent caves, perhaps on either side of a headland, meeting to form a **natural arch**, which may last perhaps 50–100 years before it collapses.

The collapsed arch can lead to development of a sea stack (although not all sea stacks are initially arches), such as the Old Man of Hoy in Orkney. Frequently, stacks are eroded away to form **stumps**. Some stacks, such as Charlie's Garden at Seaton Sluice on the Northumberland coast, have now vanished.

### Summary

- Subaerial processes include physical, chemical and biotic weathering, and mass movement such as landslides, slumps and rockfalls.
- Weathering breaks off rock fragments of varying sizes, which fall to the foot of the cliff, protecting the cliff from erosion.
- Mass movement is the downslope movement of material, aided by gravity.
- Marine erosion by waves depends on wave characteristics and the wave environment of the coastline, the geological environment and the morphology of the coastline.
- The combination of marine erosional processes and subaerial processes results in the formation of distinct coastal landforms.
- Cliff morphology is influenced by geology, especially the hardness and structure of the rock, the balance between marine erosion and subaerial processes, and inherited characteristics.
- Headland and bay formation results from the differential erosion of juxtaposed rocks of varying resistance.

# Processes of coastal transport and deposition and the associated landforms

## Modes of sediment transport

Once in motion the mode of transport that a sediment grain is subject to is largely determined by the mass (size) of the grain and the speed of the currents.

- **Bedload** is where grains are 'supported' by either continuous **traction** or intermittent contact (**saltation**) with the sea floor. When traction occurs, the grains slide along the seabed. Traction is a relatively slow mode of transport — while strong currents can transport pebbles, cobbles and boulders, weak currents only transport sand-sized particles. Only the strongest currents can move pebbles by **saltation**, where grains are bounced along the seabed. Saltation is an important mechanism for sand transport by the wind (aeolian).
- **Suspended load** is where grains are supported by turbulence, typically when moderate currents are transporting silt or clays in **suspension**.
- **Solution** involves corroded particles from limestones being dissolved by salt water and carried by the sea.

Overall, sand particles are the easiest to transport at relatively low speeds, whereas cohesive, fine clay particles and larger gravels, pebbles and cobbles require higher speed currents. Different transport speeds mean that sediment is increasingly sorted over time, through the progressive differential transport of grains of different sizes.

> **Knowledge check 11**
>
> Explain why beach deposits are generally 'sorted'.

## Longshore drift

Just as constructive waves 'swash' materials onto the beach, waves that are influenced by **prevailing winds** move material *along* the shoreline.

Longshore drift most commonly takes place when the prevailing winds blow at an oblique angle to the shoreline and therefore cause incoming waves to approach the shore at an oblique angle, whereupon the swash pushes pebbles and sand up onto the beach diagonally (Figure 13).

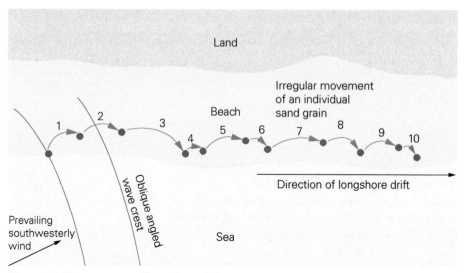

**Figure 13** The process of longshore drift

However, under the influence of gravity, the backwash from the same wave moves back towards the sea at right angles, carrying some of the deposited beach material with it. The long-term impact of these alternate swash and backwash movements is to move sand, pebbles and gravel along the shore from one end of the beach to the other.

Longshore drift is an irregular process on a day-to-day basis — its direction depends on both the direction of prevailing winds and the orientation of the shoreline. In the longer term, over months and years, longshore drift operates in a preferred direction — for example, on the UK's south-facing coasts it transports materials eastward, while on west-facing coasts the direction is northward.

Overall, longshore drift is especially marked on straight coastlines but, providing the downdrift output of sediment is matched by updrift input, the beach system remains in **equilibrium**. Only if the supply of material is disrupted is there a need to replenish and maintain beach stores (by beach feeding and groyne construction).

## Deposition

Deposition of material in open water occurs when the energy of the transporting water becomes too low to transport the sediment, with the sediment deposited being directly proportional to the mass (size) of the sediment (Stokes's law). Where clay particles aggregate together to form **flocs, flocculation** increases the fall velocity and therefore speeds up deposition. In shallow waters and beach situations the energy of the waves causes differential deposition, with the largest cobbles and boulders thrown above the high-water mark by storm waves to form a **storm beach** (as at Newgale, Pembrokeshire) — one example of sediment sorting.

## Characteristics and formation of coastal depositional landforms

### Beaches

Beaches are the most common depositional landform. They are composed of loose, unconsolidated sands and pebbles and yet, paradoxically, usually survive the roughest

**Knowledge check 12**

Why might a river dam constructed inland contribute to changes in coastal depositional landforms?

storms and most energetic wave conditions — largely because of their mobility and their ability to adjust their dynamic equilibrium to many different conditions.

## Beach forms

### Large-scale landforms

The main influence on the beach plan is wave energy — in particular, its relationship with the prevailing wave direction. The interactions between the amount of longshore movement of sediment, the prevailing wave conditions and the sediment supply result in three main types of beach.

1 **Swash-aligned beaches** occur where waves break parallel to the shore, with little longshore drift. Sediment movement is onshore–offshore.

2 **Drift-aligned beaches** occur where waves arrive at an oblique angle and considerable longshore drift occurs. Plenty of sediment is available.

3 **Zeta-formed beaches** are at an oblique angle to the dominant wave approach, with longshore drift, but where headlands at each end cause wave refraction and block sediment movement. At the far end (away from the dominant wave direction) sediment builds up in front of a headland.

## Spits

Spits are linear deposits of sand and shingle, attached to land at the **proximal** end but free at the **distal** end. They are found where:

■ the coast has an abrupt change of direction, such as at an estuary or bay, both of which are low-energy environments

■ there is a ready supply of sediment, particularly sand and shingle

■ longshore drift is active

■ tidal range is micro-tidal (less than 2 m), so wave energy is focused into a restricted zone

Many spits have recurved or hooked distal ends, which may result from wave refraction, currents or a combination of the two. Spits create sheltered areas behind them in which salt marshes frequently form (Figure 14).

> **Exam tip**
>
> Practise drawing annotated diagrams of the formation and key features of the different landforms. Research a named example of each landform.

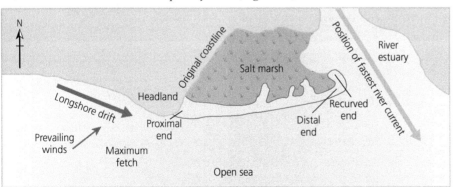

**Figure 14** The main features of a spit

Longshore drift moves the sand and shingle along the shore and, where there is a major change in the coastline trend, the longshore drift may continue to deposit sediments into the sea, gradually building up the spit. Most spits grow at fairly rapid rates. Orford Ness in Suffolk, for example, is extending southwards at a rate of up to 15 m per year and has diverted the River Alde 12 km south.

Spits can develop so that they form a bar, cutting off large areas of sea and creating coastal lagoons. The 98 km-long Neringa Spit in the Baltic Sea has virtually sealed off the lagoon behind it, so the Lithuanian port authorities have to dredge a channel through it to keep the port of Klaipeda open. A pair of spits facing each other on either side of a coastal indentation can sometimes occur, as at Christchurch Harbour in Dorset.

A spit is often in a state of precarious equilibrium between inputs of wind, wave energy, tidal energy and sediment, and is therefore subject to frequent change.

## Bars and barrier islands

**Bars** are elongated deposits of sand and shingle, usually lying parallel to the coastline, laid down by constructive waves and separated from the shore by lagoons. They may vary in scale from comparatively small features just a few metres wide and a couple of hundred metres long, such as Loe Bar, Porthleven, Cornwall, to landforms over 1 km wide, hundreds of kilometres long and up to 100 m in height. At this larger scale, bars are known as barrier beaches or **barrier islands** and, unlike some of the smaller bars, they are not submerged at high tide.

In Europe the biggest system of barrier islands is the Frisian Islands, which stretch from the northern part of the Netherlands into North Germany, enclosing a huge, shallow sea — the Wadden Sea.

Bars and barrier islands typically occur in seas with shallow offshore gradients and a low tidal range (less than 3 m) yet with relatively high wave energy.

## Tombolos

Tombolos are complex features that develop when longshore drift joins an island onto either the mainland or to a larger island. A spectacular tombolo — the tombolo di Orbetello in Tuscany — was formed where two sand spits joined the former Monte Argentario Island onto the mainland, forming a large lagoon. Some smaller tombolos on the Isles of Scilly have formed where a cuspate foreland has linked up with islands.

## Cuspate forelands

Cuspate forelands are triangular projections, with an apex pointing out to sea. They vary considerably in scale: while the Dungeness foreland extends for around 30 km along the Kent coast, projecting about 15 km into the English Channel, some of the cuspate forelands in North Carolina on the east coast of the USA, such as Cape Hatteras, extend up to 150 km along the side attached to the mainland.

Cuspate forelands seem to be present where sediment, moved by longshore drift, becomes trapped when an equilibrium is reached between sediment inputs and energy available to move it.

The Dungeness cuspate foreland (the most extensive foreland in Europe) developed from two opposing directions of swell and the consequent longshore drift.

**Exam tip**

You can research a number of theories to explain the formation of different types of bars and barriers in different locations.

**Exam tip**

Make sure you can recognise how different coastal landforms are depicted on an OS map.

## Summary

- Coastal transport processes involve solution, suspension, saltation and traction, and include longshore drift. The mode of transport for a sediment grain is determined by the size of the grain and the speed of the currents.
- Longshore drift occurs where waves, influenced by prevailing winds, move material along the shoreline.
- Processes of coastal deposition result from reduced energy levels and include flocculation and sediment sorting.
- Depositional coastal landforms include beaches, spits, bars, barrier islands, tombolos and cuspate forelands.
- Beaches adjust their dynamic equilibrium to different conditions.

# The role of aeolian, fluvial and biotic processes in the formation of coastal landforms

## Aeolian (wind) processes in sand dunes

Coastal sand dunes are common features of many coasts, especially in mid latitudes. Excellent examples are found along the Netherlands coast, Les Landes in southwest France and in parts of the UK.

Dunes develop above high-tide level and can extend several kilometres inland. Some dune systems consist of a sequence of ridges and troughs parallel to the shore, whereas others are more complex dune fields with some ridges at right angles to the sea. The height of the dune ridges varies from 1–2 m up to 30–40 m above sea level. Optimum conditions for a dune system to form are:

- an abundant supply of sand, usually sourced from the seabed
- a low beach gradient
- a macro-tidal range that exposes a large area of beach for reworking
- strong prevailing onshore winds
- an area of 'inland space' for the dunes to develop on
- vegetation, such as sea couch and marram grass, to colonise the dunes once they have formed

The initial movement of sand by wind occurs when critical wind velocity is attained relative to a given sediment particle size (threshold velocity). Once the initial sand movement occurs and the particles are entrained, transport occurs, with the most common process being **saltation**. For sand to be deposited again, a reduction in wind velocity is needed, for example in the lee of obstacles or as a result of the frictional impact of vegetation. Sand then accumulates quickly to form a streamlined dune form with a gently sloping upward (stoss) side and steeper-gradient downwind (lee) side.

### Knowledge check 13

Define saltation and explain why it is the most common means of transport in sand dune formation.

A single foredune ridge is the basic requirement for the establishment of a coastal dune system, dependent on the availability of further sediment supply from the source beach.

- If sediment supply is low then sand blown inland from the foredune may not be replaced, rendering the dune vulnerable to storm erosion and **blowouts** by wind **deflation** — loose sand may reform as **parabolic dunes** behind the blowout.
- Foredune morphology will be maintained if the loss of sand from the system is matched by new supplies from the beach.
- Where sediment supply from the beach exceeds that lost from the foredune (net sediment gain) a whole series of dune ridges may be formed. Dunes usually begin to form above the spring high tide level.

If **pioneer** plant species — i.e. those that can survive scarcity of water, mobile land surfaces, high levels of salt, exposure to strong winds and a low level of nutrients — establish themselves (such as sand twitch or sea couch grass), frictional drag from the stems and leaves of the plants reduces wind speed, allowing sand to accumulate.

Marram grass and roots also trap sand, leading to the development of small, low embryo dunes. If sufficient sand accumulates, adjacent embryo dunes merge to form a line of foredunes, which develop into sizeable ridges and are colonised by more vegetation. Migration landward of new yellow dunes occurs when sand grains move up and over the ridge and are deposited on the lee slope, moving at rates of up to 5–7 m per year. Ultimately, older, grey dunes, a long way from the sea, become fixed as they are covered by vegetation.

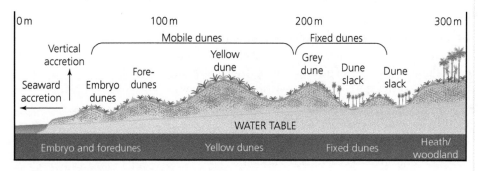

**Figure 15** Sand dune zonation

A sequence of parallel ridges can then develop, extending inland (Figure 15). Between the ridges, hollows known as **slacks** are found. As soon as the wind crosses a ridge, its speed at ground level initially falls, then increases again towards the bottom of the lee slope, so eroding the slack, which may contain water. If the wind drills down to the water table, especially in winter when rainfall is usually high, a lagoon is formed. Many dunes contain blowouts. These are initiated by the removal of significant amounts of vegetation by human and animal activity, which leads to the wind removing the sand by deflation. This is an example of a **positive feedback**:

    loss of vegetation → sand removal → increased wind speed as less friction from plants → more mobile sand → greater difficulty of vegetation re-establishment

**Knowledge check 14**

Why is the loss of vegetation and resultant changes to the dunes an example of positive feedback?

# Fluvial processes in estuarine environments

Both estuaries and deltas are locations where rivers extend into the coastal zone. They result from interaction between marine and fluvial processes, i.e. salt water and fresh water. The degree of mixing of fresh water and salt water in an estuary influences ecosystem development.

The sediments found in an estuary come from three different sources:

1. fluvial/glacial land-based sources
2. estuary margin sources
3. sources outside the estuary itself, for example from cliff erosion along the coast and longshore drift towards the estuary mouth

In the low-energy environment of an estuary the dominant process is deposition, so the estuary can be regarded as a sediment sink for sand and mud.

## Tidal flats

Around the edges of estuaries, extensive unvegetated depositional areas are found, known as **tidal mudflats**. These are intertidal areas. At low tide intricate patterns of channels and rills are exposed, formed by fresh water from tributaries flowing across the mudflat to the sea. Most mudflats are a mixture of sand and mud, for example Afon Mawddach near Barmouth on the west coast of Wales, but they do show some zoning as a result of sorting.

In terms of energy input, the outer (seaward) part of the estuary receives much tidal and wave energy, while the inner (landward) part of the estuary has considerable energy inputs from river currents. The result is that finer sediments (mud) are transported through these two areas, into the 'less energetic' central zone, where they are deposited. The coarser sands tend to be deposited in the inner and outer parts of the estuary in the more energetic zones.

Tidal mudflats provide ideal environments for organisms such as lugworms ('the fisherman's friend'), which churn up the tidal mudflats by burrowing, a process known as **bioturbation**.

## Salt marshes

Salt marshes usually exist in protected, sheltered environments, including behind spits and bars and along the fringes of estuaries, but are also found on open coastlines.

At a macro-scale, a salt marsh coastal system appears as a near-horizontal platform that slopes gently seaward, but at a micro-scale a wide variety of features, such as channels, rills and salt pans, can be seen concentrated around tidal creeks.

The platforms are built up by the deposition of sediment brought onto the marsh surface by flood tide currents, and then trapped by vegetation. Sediment **accretion** thus leads to platform elevation. As the rate of sediment accretion is greater in the lower part of the platform, the lower marsh areas are elevated at a faster rate, making the platform almost horizontal over time.

Like sand dunes, salt marshes result from the interaction of geomorphological and ecological processes. The key to their formation is a low-energy tidal environment, where the sea's erosional ability is limited, permitting plant colonisation and sediment accretion.

The 'Saltmarsh Coast', which contains many examples of salt marsh, is found along 120 km of Essex coast between the River Crouch and Blackwater estuaries. Salt marshes are both complex and fragile. Like many other coastal sedimentary systems, salt marshes are extremely sensitive to changing environmental conditions that affect the rates of sediment accretion or the level of erosive processes. The environmental changes may be natural, such as the impact of rising levels or storms, or human induced, such as the impact of draining and land reclamation, or the accidental introduction of spartina grass (an invasive or alien species). Salt marshes are threatened by a huge range of factors, but they are so valuable in the goods and services they provide that there are many management plans in place to conserve them. As a result of a coastal management strategy of managed realignment (page 50), new salt marshes are being artificially created.

## Biotic processes in mangroves and coral reefs

### Mangroves

The principal difference between salt marshes and mangroves is the greater above-ground biomass in mangroves. There are many different species of mangrove tree, all of which can tolerate relatively high levels of salt in the water. They are restricted to a zone of about 30° either side of the Equator. Mangroves have multiple aerial tap roots that emerge from the trunk, above the mud, which anchor the tree, help with oxygen uptake and assist in the trapping of sediment (so mangroves are an important sediment store). The resulting increases in deposition and vegetation can play an important role in reducing the impact of wave energy on the coastline. The obstacles caused by aerial roots and low branches can result in a reduction in wave height of up to 66% over a 100 m width of mangrove. Mangroves can provide 800% more coastal protection and 470% more erosion prevention than salt marshes. As occurs on tidal mudflats, **flocculation** of fine particles is a key process. In a mangrove forest (mangal), zonation occurs.

### Coral reefs

A coral reef coast is an example of a rocky coast. Coral coastlines are found in about 100 countries and are grouped into three main formations geographically: the Indo-Pacific formation, the Western Atlantic formation and the smaller Red Sea formation.

Coral reefs build up over time, as each coral polyp (a small animal with tiny tentacles) secretes a calcareous skeleton that it leaves behind when it dies. Growth rates are usually 1–100 cm per year and reefs eventually achieve a thickness of hundreds

**Knowledge check 15**

Draw a spider diagram to summarise the physical and human factors that affect salt marsh development.

**Exam tip**

Obtain a map of the global distribution of both mangroves and coral reefs so that you can locate named areas.

of metres. Coral polyps have a symbiotic relationship with tiny algae known as **zooxanthellae**. The ideal conditions for coral growth are:

- a clear sea, to allow light penetration (avoidance of silted areas such as river mouths)
- tropical water temperatures of between 23°C and 29°C
- sea salinity of between 30 ppt and 40 ppt
- shallow water no deeper than 100 m
- well-aerated water, resulting from relatively strong wave activity

Dead coral is susceptible to erosion by wave activity, which can reduce the coral limestone to rubble. This can be transported by tidal currents to infill active reef structures, to create rubble mounds suitable for new coral colonisation, or to accumulate in larger piles to create coral islands (**cays** and motus).

The morphology of reefs usually comprises a 'fetch-facing' outer windward side, which is eroded by the breaking waves, and a relatively protected 'leeward' side, where new coral colonies are developing. Debris from the eroded windward side is often deposited on the 'platform' surface, graded from coarse to fine debris towards the lee side. Wave refraction contributes to the piling up of debris to begin the process of cay formation.

Coral reefs can act as a first line of defence by being a barrier to storms, reducing the impact on a coastline by decreasing the height of incoming waves by 84% and wave energy by 97%.

## Summary

- Non-marine influences in coastal environments include wind action, river action and biotic processes.
- Sand dune formation requires a supply of sand, a low beach gradient, a large area of exposed beach, strong prevailing onshore winds, an area for the dunes to develop in and vegetation to colonise the dunes once they have formed.
- Sand dunes begin to be formed when wind speed is sufficient to move sediment of a given particle size. Particles are commonly transported by saltation and deposited once the wind speed drops.
- Fluvial processes in estuarine environments form tidal flats, salt marshes and micro-features such as channels and rills.
- Tidal flats are extensive, unvegetated intertidal areas. Salt marsh coastal systems form a near-horizontal platform sloping gently seaward. They are built up by the deposition of sediment (accretion) brought by flood tide currents and trapped by vegetation in a process of succession.
- Biotic processes influence the development of coral reefs and mangrove coastlines in mostly tropical locations.

# Variations in coastal processes, landforms and landscapes over different timescales

Table 5 shows the scale of coastal changes in relation to both absolute and human timescales.

**Table 5** The scale of coastal changes over time

| Absolute timescale | Human timescale | Coastal process |
|---|---|---|
| Seconds<br>Minutes<br>Hours<br>Days<br>Weeks<br>Months<br>Years | Dumping of litter, sewage<br>Emergency defences against erosion and floods<br>Impacts of tourists, visitors and local population<br>Coastal management decisions<br>Coastal development | Sediment movement by wind or water<br>Cliff falls from mass movements<br>Tidal cycles, shore normal sediment movement and storms<br>Storm surges, breaks of defences<br>Beach scour<br>Shore profile adjustment<br>Tidal cycles<br>Shore profile adjustment (seasonal)<br>Coast accretion erosion<br>Coastal process response to defences<br>Longshore drift |
| Decades | Coastal defences | Erosion and accretion cycles<br>Coastal process response to defences<br>Formation and loss of habitat |
| Centuries | Shifts in settlement | Historic coastal development, loss of towns and villages to the sea |
| Millennia | — | Sea level changes in response to glaciation, tectonics etc. |

*Increasing timescale* (downward arrow)

## Rapid process and landform change

High-energy storm events, the resulting erosion of beaches and dunes and the rapid mass movement processes in cliffs can take place over a very short time span — from seconds to hours, and certainly over a day.

### Storm-generated erosion

Storm-generated erosion can occur on very short timescales. At midday on Monday 26 February 1990, a violent storm breached the sea wall at Towyn, North Wales, flooding 10 km² of low-lying land, and damaging 2800 houses and 6000 caravans.

The Towyn floods were caused by a range of **physical factors**. A particularly deep depression caused a storm surge 1.3 m high that coincided with extremely high spring tides and strong 130 km h⁻¹ onshore winds, leading to overtopping of the sea wall (a 1-in-200-year event). Huge volumes of seawater spilled through the gap, flooding the pumping stations used to drain the land used for farming. **Human factors**, such as failure to repair the sea wall and the fact that the embankment had not been improved to be 'future proof' against rising sea levels, exacerbated the floods. Usually, new development triggers sea wall defence improvements. However, although there had been recent large-scale development of retirement bungalows and holiday

> **Exam tip**
>
> Always keep your case studies up to date using the internet.

arks, supported by planning permission, on these very low-lying areas at Towyn, the planned upgrading of sea defences had not been carried out. The actual breach of the sea wall happened over a period of around an hour, and flooding occurred rapidly over a wide area, augmented by river flooding.

Many smaller-scale, high-energy storm events have occurred since the Towyn flood. For example, in March 2018 a storm resulted in erosion of a 6 m-high dune near Hemsby in Norfolk. The foundations of some houses were exposed and homeowners lost over 5 m of garden, with 10 homes evacuated.

Worldwide, much of the severe coastal damage comes from hurricanes, typhoons and cyclones, for example the tropical storms Hurricane Katrina on the US Gulf Coast (2005) and Typhoon Haiyan in the Philippines (2013). The impact of a tropical storm is felt over a number of days as the storm migrates, with any one area usually affected for 1–3 days.

Rapid mass movement processes include rockfalls, landslides and slumps. In July 2012 a 22-year-old woman was buried by 400 tonnes of rock when a section of sandstone and clay cliff collapsed onto a beach at Burton Bradstock, Dorset. In August 2018 a 9-year-old girl died at Staithes, Yorkshire when rocks fell from the cliff.

## Changes at a seasonal temporal scale

There is a strong relationship between waves (in particular, wave steepness) and the angle of beach profile. Wave steepness is the ratio of wave height to wavelength — the higher the value, the greater the energy brought by the wave onto the beach.

As Figure 16 shows, there are seasonal changes in the angle of the beach profile. The beach is in short-term equilibrium with the contrasting seasonal conditions.

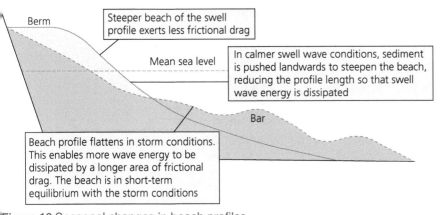

Berm

Steeper beach of the swell profile exerts less frictional drag

Mean sea level

In calmer swell wave conditions, sediment is pushed landwards to steepen the beach, reducing the profile length so that swell wave energy is dissipated

Bar

Beach profile flattens in storm conditions. This enables more wave energy to be dissipated by a longer area of frictional drag. The beach is in short-term equilibrium with the storm conditions

Figure 16 Seasonal changes in beach profiles

In winter the waves tend to have more energy (generally stronger winds), so they erode and transport sediment offshore, possibly forming an offshore bar, and thus lowering the beach profile. During summer the waves are generally less energetic, so they move material onshore, building the beach up.

**Knowledge check 16**

Explain how the Towyn floods resulted from a combination of both physical *and* human factors.

**Exam tip**

Storm surges also prove useful case studies, such as those associated with deep depressions in the North Sea in the winter of 1953, January 1978 and December 2013. If storm surges are combined with high tides and strong onshore winds, they represent a major threat to sea defences. Search www.surgewatch.org for information on storm surges affecting the UK.

Tides also have an impact on beach profile. In areas of monsoonal or other markedly seasonal climate in terms of rainfall, the wet and dry seasons show clear beach profile changes too.

A further complication is the influence of the beach steepness on the waves. On a steep beach waves tend to be plunging, and a significant amount of the incoming energy is reflected back from the beach (known as a **reflective** beach). Conversely, on a shallow-angled beach the waves break and spill as the wave base is reached further out from the shore, thus dissipating the wave energy as the waves move across the wide beach (known as a **dissipative** beach).

The calibre (size) of the beach material can also affect the angle of steepness, with steeper beaches being associated with larger sediment and shallower beaches being associated with fine-grained sediment. The link between sediment size and beach gradient is thus a result of contrasting **percolation** rates. Moreover, particle size affects the angle of rest — the larger the particle size, the steeper the angle of rest.

# Process and landform changes over millennia

Sea level is continuously changing. Globally, all coasts have undergone considerable variations in sea level over geological time and especially since the Quaternary period, beginning about 2 million years ago. Some coastlines have experienced more 'recent' sea level change than others. Since the last extensive glaciation, sea levels have risen by about 120 m.

Changes in sea level alter energy inputs and outputs and are therefore important for the development of landforms.

## Causes of sea level change

**Eustatic** changes involve changes in **absolute** sea levels and are **global**, because all oceans and seas are interconnected. While the total volume of water in the global hydrological cycle (a closed system) is constant, changes in where and how water is stored cause eustatic change.

Eustatic changes are therefore unrelated to local/regional effects. They are most commonly caused by changes in the ocean water volume and temperature (thermal expansion and contraction of the ocean). As oceans cool, their volume contracts and when they warm it expands. While this is a worldwide change, the impact is more marked in tropical oceans. Similarly, a decrease in salinity causes a rise in sea level, and vice versa. These changes in salinity and temperature are believed to make a difference of only about 10 m to global sea levels, and so are of much less significant than changes in sea level associated with the accumulation or melting of ice. These are **glacio-eustatic changes**, which occur when ice on continental ice caps and ice sheets forms and melts, removing water or releasing it back into the oceans. (Note that the increase and decline in the size and amount of floating sea ice, including icebergs, ice shelves and ice packs, has *no* influence on sea level. This is because ice displaces the equivalent of its own mass.)

It is estimated that if the Greenland and Antarctic ice sheets melted there would be a eustatic sea level rise of up to 90 m. However, sinking of the ocean basins under the influence of the weight of all the water (hydro-isostasy) would diminish the impact (negative feedback) to around 60 m.

**Isostatic** sea level changes are **regional** changes of sea level. The Earth's crust 'floats' on a denser underlying layer (asthenosphere). This two-layer system is in isostatic balance when the total weight of the crust is exactly balanced by its buoyancy. The addition of a load at a particular point of the crust (which could be water, ice or increasing sediment from a large delta) upsets this equilibrium. To compensate for the increased coastal weight, some of the asthenosphere flows away, causing isostatic depression as the land level falls.

**Isostatic readjustment** or **rebound** occurs, for example, when the ice sheet melts and the land reverts back to its former position.

Figure 17 shows the complex relationship between sea level changes and their impact on landforms.

**Knowledge check 17**

Distinguish between eustatic and isostatic changes in sea level.

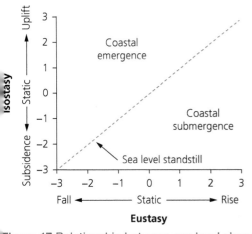

**Figure 17** Relationship between sea level changes and landforms

**Glacio-isostasy** occurs when glaciated continents are depressed by the weight of their glaciers. During glacial periods ocean water volumes and global sea levels are low, but glaciated continents are regionally depressed by the weight of glaciers, so their **relative** sea level is higher than that experienced on non-glaciated continents. Upon deglaciation, the removal of the weight of ice causes isostatic rebound or recovery of the land, as is apparent along the coasts of Scotland and the Gulf of Bothnia, between Finland and Sweden.

**Hydro-isostasy** occurs when the volume of water released into an ocean basin exerts a weight on the ocean floor and depresses it by an amount roughly equivalent to one-third of the depth of the additional water.

In terms of landform formation, sea level changes can be classified as **negative** where there is a relative fall in sea level (i.e. a marine regression). An **emergent** coastline then results, with the coastline building out from its previous position. There are three possible scenarios for a relative fall in sea level.

1 The sea level falls and the land either rises, stays still or subsides at a slower pace.
2 The sea level remains fixed while the land rises.

**Exam tip**

This is a complex area. Try to understand it using simple diagrams. Think of relative sea level rise and positive and negative impacts.

**Exam tip**

For the Eduqas or WJEC exams you only need to learn about *either* isostatic *or* eustatic changes in sea level, but knowledge of both is relevant for understanding the resulting formation of landforms.

3  The sea level rises but the land rises at a greater pace, for example in northwest Scotland, which is experiencing a postglacial rise in sea level, but with a rapid isostatic recovery.

**Positive** sea level change occurs where there is a relative rise in sea level (i.e. a marine transgression) and a **submerged** coastline results, leading to drowning of the coast and the inshore migration of some landforms, for example beaches. For a relative rise in sea level there are again three possible scenarios.

1  The sea level rises and the land either subsides, stays still or rises at a slower pace.

2  The sea level remains fixed while the land subsides.

3  The sea level falls but the land subsides at a greater pace.

A number of factors influence the landforms produced by changes in sea level as well as the sea level change itself:

- the structure of the coast, whether concordant or discordant (page 25)
- the relief of the coast, whether highland or lowland
- special factors, such as whether the coast was glaciated or not before the sea level change

## Emergent landforms

Emerged coastlines are evident when beach deposits and marine shells, for example the patella (limpet) shells on the raised beaches along the Gower Coast, are found stranded above the present-day high-tide level, creating a **raised beach** that is often backed by **relict** or fossil cliffs. These beach deposits and shell beds were formed when the sea was at a previously higher level.

The raised shore platform creates a **marine terrace**, an area of flat land that can be utilised for farming, as the soils are relatively sandy, or for communications. The presence of a raised platform indicates that, before the relative fall in sea level, the sea must have been at a higher level for a considerable period of time to have allowed the shore platform to develop. Relict cliffs can be identified by a marked break in slope, but they can show signs of degradation from subaerial processes.

Sometimes, as in western Scotland, where raised beaches have been identified at 8 m, 15 m and 30 m above present-day sea level, there are **series** of terraces and relict cliffs. An analysis of the raised beach deposits shows how they have been rounded by corrasion and attrition. Sometimes, fossil sand dunes can be found high above current sea levels. Associated with the relict cliffs, it is possible to find relict caves (e.g. King's Cave in Arran or Paviland Cave in Gower) and sea stacks.

The extent to which raised beaches and their landforms remain intact often depends on their geology and subsequent level of resistance to denudation, as well as their age.

# Submergent landforms

Positive changes in sea level mean that the overall shape of the coastline is changed as river and glaciated valleys and coasts are drowned, creating an indented coast of inlets (e.g. rias and fjords), bays and promontories. The sea may also isolate some areas of land to form islands. For example, Great Britain became physically separated from Europe during the Holocene Flandrian transgression.

## Rias

**Rias** are more common along **discordant** coastlines, where geological strata trend at right angles to non-glaciated coastlines. When global sea levels were lower than at present, base level was lowered, giving rivers a renewed energy to cut downward — forming deep, rejuvenated river valleys. Once sea levels began to rise again, drowned river valleys, with their winding dendritic pattern, were formed. Rias are tidal, and their central deep-water channels are of vital importance for ports, an example being the oil/gas terminals at Milford Haven in southwest Wales (Figure 18).

(a)

(b)

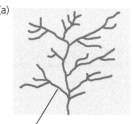

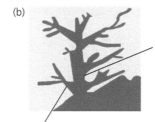

River channels become wider and deeper with submergence

Before submergence the river channel network shows a typical dendritic pattern

The channel network is partially submerged

**Figure 18** A typical ria (a) 18,000 years BP, (b) present day

Rias are found around the world. Other examples include those of southern Cornwall and Devon, such as the Kingsbridge estuary, Devon, as well as Galicia on the Iberian Peninsula.

Features similar to rias are formed on **concordant** coastlines where geological strata, mountains, valleys and rivers all trend parallel to the shore, for example on the Dalmatian coast of Croatia. Wide, open bodies of water called **sounds** develop, which have longitudinal islands and ridges of raised land between them.

## Fjords

**Fjords** (or fiords), sometimes known as sea lochs, are **drowned glacial valleys** that have been shaped by the action of ice and submerged during the Holocene. The deep, steep-sided troughs carved out by moving ice are submerged as sea levels rise. Fjords are formed at higher altitudes, where the effects of ice have been more profound. In some upland coastal areas they can reach in excess of 1200 m in depth.

Fjords have steep sides and flat bottoms, typical of the U-shaped valleys of glaciated landscapes. They are relatively straight-sided and narrow compared with rias.

Also typical of glaciated areas, fjords may have hanging valleys and waterfalls. Ridges of scree and moraine may also line the fjords' shores, providing evidence of their glacial heritage. Fjords have a shallow entrance called a **threshold**. The threshold is occasionally less than 100 m in depth. If the threshold appears at the water's surface it may manifest as islands called skerries (a **skerry** is a small rocky outcrop). The threshold could be a large terminal moraine — material 'bulldozed' by the moving glacier, marking its furthest point of advance during the glacial period — or it could be the point where the glacier's snout was thinner, or perhaps where the glacier began to float, giving the glacier reduced erosive power (Figure 19).

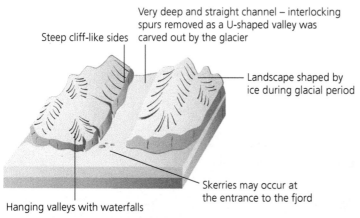

Very deep and straight channel – interlocking spurs removed as a U-shaped valley was carved out by the glacier

Steep cliff-like sides

Landscape shaped by ice during glacial period

Skerries may occur at the entrance to the fjord

Hanging valleys with waterfalls

**Figure 19** The key features of fjords after submergence

Examples of fjords can often be seen in high-latitude locations, such as British Columbia in Canada, southern Chile and western Scotland. The Sognefjord in Norway is over 200 km long and reaches a depth of 1300 m. Many of the islands found off the coast of Norway are skerries, linked to the 1190 fjords found there.

## Summary

- High-energy storm events, the resulting erosion of beaches and dunes, and rapid mass movement processes in cliffs can take place in a very short time span, from seconds to hours or over a day.
- In winter, waves tend to have more energy because of stronger winds, so they erode and transport sediment offshore, lowering the beach profile. During summer, waves are less energetic, so they move material onshore, building the beach into a steeper profile.
- Changes in sea level alter energy inputs and outputs and influence the development of landforms.
- Eustatic changes involve changes in absolute sea levels and are global, because all oceans and seas are interconnected. In contrast, isostatic sea level changes are regional and caused by movements of particular parts of the Earth's crust.

**Knowledge check 18**

Explain how you would distinguish between a ria and a fjord.

**Exam tip**

For the Eduqas and WJEC specifications you must select *one* landform and explain the impact of changes of sea level on its formation. Rias, fjords or raised beaches would all be a good choice.

# Coastal processes are a vital context for human activity

## Positive impacts of coastal processes on human activity

The coastal zone attracts a range of human activities, especially landward of the high-tide level, where low-lying, flat land in a pleasant environment has the potential for many land uses. Intertidal and offshore zones attract other activities, such as tourism, fishing and port development.

Of the inhabited continents, only in Africa do more people live in the interior than in the coastal zone. In Asian countries over 1.5 billion people live within 100 km of the sea. By 2025, about 75% of the residents of the USA are expected to live in coastal areas, which together only account for 17% of the land area, with Florida and California leading the way. In many parts of the world the coast is a very crowded zone — for example in China, where coastal population densities average 600 persons km$^{-2}$. In the USA, China and most of Latin America more than 70% of major cities are coastally located.

## The growth of tourism

Over the past 200 years tourism and recreation have developed into major economic activities along the coast.

By the end of the nineteenth century, in most developed countries, seaside holidays were part of the annual rhythm of people's lives, hence the rise of the British seaside resort. As transport has become increasingly sophisticated and journey times dramatically reduced, global travel has become a reality for large numbers of people, as they acquire both the time and money to travel, and coastal areas have played a significant role in this global explosion of tourism.

There are certain key resources necessary for the development of coastal resources for tourism and recreation.

**Physical resources** include attractive coastal scenery (e.g. cliffs) and ecosystems (e.g. coral reefs), high-quality seawater that is free of strong currents and at a pleasant temperature, and sunny climates. Different combinations of these physical resources can result in different types of tourism. Well-known examples of world-class beaches include those in Florida and the Mediterranean. The physical resources of the Pembrokeshire Coast National Park in South Wales attract over 7 million day visits a year.

As well as physical resources, tourism also needs **human resources**, for example cultural attractions, heritage resources, quality provision of services, accessibility and opportunities for recreational activities. Dubai is an example of a coastally located area with many artificial attractions, which have helped make it one of the fastest-growing tourist destinations in the world.

While there are numerous references to the negative environmental impacts of tourism, these have to be counterbalanced by the positive impacts, both economic and social.

> **Exam tip**
>
> Research examples of coastal tourism. In each case, which factors have encouraged its growth and what impact has it had?

Coastal tourism aids economic development in a number of ways. Foreign exchange earnings can be enhanced by levying taxes, such as tourist and airport tax. For developing nations, construction of hotels and resorts has such high capital costs that it can only happen if there is inward investment (foreign direct investment (FDI)) from countries such as China, the UK and Japan. The trend in tourism, as in other service industries, is towards greater **globalisation** of operations. Hence, large transnational corporations (TNCs) are increasingly involved, often acting as conglomerates, with shares in hotels, airlines and leisure developments.

Tourism is the world's largest employer, with an enormous range of jobs available. The EU has estimated that coastal tourism has created 3.2 million jobs, generating over €180 billion. There is a small core labour force of full-time highly paid professionals, supported by a large number of workers, often poorly paid, who are engaged in low-cost, labour-intensive functions, such as cleaning. Often, the better jobs do not go to local people, but this is being remedied by training programmes. There are also plentiful opportunities for local entrepreneurs, for example in craft industries or personal services such as guiding. The diversity of employment opportunities acts as a magnet for local people, for example Ecuadoreans drawn to the opportunities provided by tourism in the Galapagos.

Tourism also generates its own **multiplier effects**. Essentially, as a resort develops, often in a tourism enclave, so does the local economy. As profits from tourism increase and become more widespread, so they begin to **trickle down** into the local economy. In theory, the multiplier effect should lead to the emergence of more local suppliers and a decreasing reliance on foreign imports.

The multiplier effect can improve the quality of local services for local people, as the people employed in tourism have money to spend or to invest. For example, in some islands in the Maldives, restaurant workers are investing in small hotels that employ both local workers and migrant labour from Bangladesh, at both the construction and operation stages.

The strength of the multiplier effect and the magnitude of its economic impact vary according to a range of factors:

- the **level of development** of the local economy — what it can supply for the expansion of tourism (goods and services) and to meet tourists' needs
- the **type of tourism** — in theory, elite tourists should spend more, but this is not the case if they are on **all-inclusive** deals
- the **organisation of tourism** — cruise ships are a contentious issue as it is not clear how much this type of tourism puts back into the local economy
- the **level of leakage**, i.e. losses of income to foreign countries — for example, when brandy is imported or multinational hotels send all their profits back to headquarters in a developed country

Whether the impacts of tourism are really positive is contentious — environmentally, economically and socio-culturally. It can be argued that the investment in people and places is positive because money earned can be used to improve both physical and human coastal environments, for example by replanting coastal wetlands or recharging the sand on beaches, as well as providing high-quality facilities. So much

**Knowledge check 19**

Apart from tourism, how else can coastal processes have a positive impact on human activity?

depends on the scale and pace of tourism development, and how well the tourism is managed to take care of the environment and support local culture.

# Negative impacts of coastal processes on human activity

A number of factors influence the rate of erosion on all types of coast. Erosion rarely occurs at a steady rate, and usually happens dramatically after a severe storm. These major events, often with a recurrence level of 1 in 100, can dramatically upset the dynamic equilibrium and lead to extensive cliff retreat (page 23) or massive beach erosion. Eustatic rises in sea level associated with global warming and the melting of ice sheets will mean that the impacts of storms are increasingly felt further inland.

**Knowledge check 20**

Explain what is meant by a recurrence level of 1 in 100.

Coastal erosion can be exacerbated by inappropriate coastal development too near to the shore or by coastal management. Many sites of dramatic change occur downdrift from newly constructed jetties or hard engineering defences, which cause sediment starvation.

The higher the value and/or density of the land development, the greater the economic losses associated with coastal erosion. **Cost–benefit** analysis is key to understanding what action, if any, should be taken to manage coastal erosion.

The overall economic and social impacts of coastal erosion are generally negative, and include:

- danger to life, in particular from sudden landslides and rockfalls from cliffs, for example along the Dorset coast
- structural damage to buildings and infrastructure, for example along the Holderness coast of Yorkshire
- damage and destruction to lifeline infrastructure, such as water, sewerage and gas pipes, as on the Louisiana coast in the USA
- loss of land that could be used profitably for farming, for example Sue Earl's Farm on the Holderness coast
- loss of vulnerable ecosystems (e.g. in the Louisiana wetlands) that have high recreational and tourism value (e.g. as bird reserves), and impacts on valuable coastal fisheries
- local instability of nearby areas, which can result in huge declines in property values for both residential and holiday amenities, as well as psychological stress for residents and owners who face financial ruin — owners may have to accept **red lining** strategies that restrict development seaward of, or near, the red line (e.g. the Norfolk coast) — with assets becoming uninsurable
- beach erosion leads to loss of beach amenity, a major problem for many holiday resorts as the lack of a beach will have huge knock-on effects on their ability to profit from tourism

The British Geological Survey estimates that, at present, 113,000 residential and 9000 commercial properties, along with 5000 ha of farmland, with a total value of £7.7 billion, are at risk from coastal erosion. This figure will increase as sea levels rise. Globally, damage from flooding due to sea level rise could be valued at over $1 trillion by 2050.

**Exam tip**

While case studies are important, do not just write a descriptive narrative. Adapt them to support a line of argument.

# Strategies to manage the impacts of coastal processes on human activity

The concept of protecting the coast is a contested one. Contrasting opinions arise — from the concerns of industry, tourism and residential sectors who see a need to 'hold the sea in place' and prevent it encroaching on the land, through to environmentalists and coastal planners who favour leaving the coast in as natural a state as possible.

There are five generic strategies for coastal management:

1 **Do nothing**, with no protection, which would lead to eventual abandonment.

2 **Managed retreat** or **realignment**, which plans for retreat using engineering solutions that recognise the natural processes of adjustment and yet identify a new line of defence.

3 **Hold the line**, which favours shoreline protection by defence from hard engineering structures such as sea walls, often supported by other methods such as groynes or beach nourishment.

4 **Move seawards** or **advance** by constructing new defences, such as offshore breakwaters seaward of original defences.

5 **Limited intervention**, which involves accommodation, by which adjustments are made to allow coping with inundation from rising sea levels by raising coastal land and building vertically, combined with state-sponsored insurance at affordable costs.

The choice of strategy or strategies is site-specific and depends on a number of factors.

**Physical factors** concern the nature of the coast itself:

- the coastal geomorphology, such as the hardness of the rock and the amount of sediment supplied by erosion (e.g. from soft rocks such as sand and clay)
- the degree of dynamism in the coastal environment, such as erosion rates, the pattern of sea level change and the precise nature of the coastal problem to be managed (whether flooding or erosion)
- the quality of the natural coastal environment, for example high-value ecosystems with rare flora and fauna (in some scenic areas hard engineering defences would be an eyesore)

There are also **financial** and **political considerations**. Of fundamental importance is the principle of **cost–benefit analysis**, as there is a need to balance the high overall costs of coastal defences against the benefits of protection, which depends on the calculated value of all the coastal development. In economic terms, expensive hard engineering defences are only worthwhile to protect and defend areas of high-value coastal developments.

The actual costs of coastal defences depend on the local characteristics of the coastal area and the type of management that is adopted. Examples of costs of coastal defences are summarised in Table 6.

**Knowledge check 21**

Explain the circumstances under which the rare option of 'advance the existing coastline seawards' might be taken.

**Knowledge check 22**

Why is it important to consider coastal sediment cells when deciding on a coastal management strategy?

Table 6 Approximate costs of coastal defences, 2019

| Coastal defence | Description | Approximate cost |
|---|---|---|
| Sea walls | Concrete structure at the cliff foot to absorb wave energy | £5500+ per metre for basic type |
| Timber revetment | Wooden structure at cliff foot to absorb wave energy while allowing some sediment flow | £1600 per metre |
| Rip rap/rock armour | Large rocks placed to absorb and dissipate wave energy | £1100–3500 per metre |
| Gabions | Tall cages filled with small rocks to add strength to the coastline | £50–60 per m³ (cubic metre) |
| Timber groyne | Structures jutting into the sea to trap longshore drift | £1100 per metre, average length 100 m, cost per groyne = £110,000+ |
| Beach nourishment | Adding sand/shingle to extend a beach or replace eroded sediment | £10+ per m³ — thousands of tonnes may be needed for most large schemes |
| Offshore breakwater | Rock island built offshore to absorb energy of waves before they reach the coastline | Millions of pounds — usually £10–30+ million |

In **political** terms there are tensions between national government's overall budget spending and local government. For many coastal communities coastal defence is seen as imperative at all costs as, ultimately, without coastal defences settlements will be inundated by storm surges or washed away by erosion.

Until recently, in many coastal areas there has been limited expenditure on scientific research, so the dynamics of coastal processes have been imperfectly understood.

**Socio-cultural factors** are also significant. Surveys have shown that many coastal residents and business owners do not feel safe unless they are protected by obvious sea wall-style defences.

There are three *broad* management strategies that can be used:

1 cliff face strategies

2 cliff foot strategies

3 beach management strategies

How site-specific factors can influence the choice of strategy can be seen in the management at Bacton on the north Norfolk coast.

The Bacton Gas Terminal was built in 1968 over 100 m inland. Coastal erosion of the soft sand and clay cliffs has resulted in parts of the terminal being only 10 m from the cliff edge today. The terminal is of national importance because it provides one-third of the UK's gas and can import and export gas to and from Europe. In 2018, £300 million was invested in updating the plant.

Hard engineering solutions could not be adopted as this would speed up the rate of erosion further along the coast, increasing the risk to settlements. The adopted solution was sandscaping. In August 2019, 1.8 million cubic metres of sand were

**Exam tip**

Note that for WJEC and Eduqas you have to select *one* management strategy to manage the impacts of coastal processes on human activity.

**Exam tip**

Beware of discussing just one coastal defence, such as building sea walls, as this gives your knowledge an overly narrow base. Another way of tackling this topic would be to research either hard or soft management strategies, and then select one of the five generic shoreline management options, such as 'hold the line' (page 46).

added to the coastline, creating a dune 6 km long, 7 m high and stretching 250 m out to sea. Wind, waves and tides would then move the sand to redistribute it, creating a beach similar to how it was 30 years ago. The scheme cost around £20 million. The Bacton Gas Terminal operators funded £14 million, with £5 million coming from the Environment Agency and £0.5 million from North Norfolk District Council. The sandscaping should provide protection of the terminal for 15–20 years. As the sand is distributed along the coastline it will increase protection for other settlements, as well as creating a recreational resource.

## Case study

### One management strategy to manage the impacts of coastal processes on human activity

#### Controlling subaerial erosion to prevent cliff retreat (cliff face strategies)

Cliffs vary in their lithology and structure and so fail in many different ways. In an ideal situation it is preferable not to defend cliffs as the sediment they supply is a vital component of the sediment system. However, people who live on top of cliffs feel entitled to some form of protection against cliff recession, which can average up to 20 m per year for cliffs of unconsolidated sands.

One future-proofing system is to use the concept of **red lining**, where the rate of cliff retreat is modelled, and any development seaward of a specified line is prevented. The coastline in the USA is classified according to its modelled 'life expectancy' and this controls decisions on development.

Cliff failure tends to be episodic, with dramatic changes after winter storms, which can combine with high tides and storm surges, to result in various types of mass movement (rockfalls, rotational slippage, mud slides/flows, and toppling — page 21). A number of strategies exist, which are frequently combined with cliff foot strategies for protection of the cliff toe.

- **Pinning** involves inserting bolts or pins through the likely shear planes, as has been done at the White Cliffs of Dover, where the cliffs themselves are a World Heritage Site.
- **Grading** involves decreasing both the height and slope angle of the cliffs to reduce the threat of mass movement and to stabilise the cliff face.

At Llantwit Major in South Wales, rockfalls from horizontally bedded limestones underlain by relatively weak shales were causing danger to recreational users. The upper cliff was blasted in 1969 to reduce its slope angle, and the blast material used as armouring to protect the cliff toe. However, the blasting weakened the rock, allowing increased weathering, and the cliff remained geologically unstable, so the scheme was considered unsuccessful.

- **Cliff drainage** is a common practice, especially for cliffs with a high clay content. Pore water pressures can be reduced by drainage lines in the cliff face, field drains, gravel trenches and by intercepting overland flow. However, this can result in subsidence of cliff-top land as the cliff dries out, and it can also have an ecological impact.
- **Gabions** are used to stabilise the cliff toe and help cliff drainage, especially behind a sea wall.
- **Vegetation** planting and fencing cliff faces can increase slope stability after grading. At Bournemouth the sand and clay cliffs have been planted with privet hedges and shrubs, and seeded with grasses and other ground cover. This improves the appearance of the cliff face and contributes to the long-term stabilisation of the regraded slope.
- **Toe protection** by sea walls, revetments/ rock armour and beach feeding often works in tandem with cliff face strategies, especially where the land is of high value or the problem is severe.

Table 7 summarises the benefits and problems of cliff face strategies. Recommended use of such strategies is restricted to areas of high risk, where alternative sediment supplies are available to compensate for losses from cliffs, and where cliff-top development is too valuable to allow cliff failure.

Table 7 Benefits and problems of cliff face strategies

| Benefits | Problems |
| --- | --- |
| ■ Increased security for cliff-top developments<br>■ Increased security for beach users<br>■ Increased security for coastal towns<br>■ Can be environmentally attractive, especially where vegetated<br>■ Avoids compensation issues that would be caused by 'do nothing' | ■ Reduced sediment supply from subaerial sources to coastal sediment budget<br>■ Reduced exposure for scientific study (geology/palaeontology etc.)<br>■ Costs do not always justify the results, which must always be supported by detailed research of marine and subaerial processes |

## Summary

- Positive impacts of coastal processes on human activity include the attractive potential of low-lying, flat land and a pleasant environment for settlement and land uses such as agriculture, transport and industry.
- Tourism, often located in coastal environments, requires both physical and human resources and is now a major global industry.
- The impact of storms and coastal erosion have a negative effect, causing social and economic losses, such as danger to life, structural damage to buildings and infrastructure, loss of land for farming, loss of ecosystems with high recreational and tourism value, impacts on valuable coastal fisheries, instability of nearby areas, psychological stress for residents, and loss of beach amenities and tourism.
- Strategies to manage the impacts of coastal processes on human activity include: do nothing, managed retreat or realignment, hold the line, move seawards or advance by constructing new defences, and limited intervention.

# ■ The impact of human activity on coastal landscape systems

## Positive impacts of human activity

Human activities, if managed effectively, can have a positive impact on coastal landscape systems. After generations of hard engineering schemes, there has been a move towards softer engineering solutions, which aim to work with the sea as opposed to against it. This section explores managed realignment as an example of how coastal management can have a generally positive impact, with ecological and environmental benefits outweighing some of the problems.

There has also been a rethink of coastal management, with the development of **integrated shoreline management** strategies where the coast is considered

**Exam tip**

Always support the strategies you describe with detailed examples.

**Knowledge check 23**

What is meant by a holistic approach to management?

holistically and managed sustainably, with humans very much part of the system (page 54).

Moreover, in response to an enormous range of pressures on the coast, a number of strategies have been developed to protect the coast and to adopt conservation management plans. Conservation strategies range from total protection through to various forms of sustainable management that allow public access.

Protection strategies operate at a range of scales and have had a positive impact:

- Global frameworks, such as the development of coastal World Heritage Sites (e.g. parts of the Great Barrier Reef, Australia or the Jurassic Coast of Dorset).
- Establishment of marine reserves at a national scale using an international (World Resources Institute) framework. In England and Wales these are called Marine Conservation Zones, and include Skerries Bank and Surrounds in Devon.
- Development of Sites of Special Scientific Interest (SSSIs), such as sand dunes, mangroves or salt marshes at a local scale. The Sefton Coast, including the Ainsdale Sand Dunes Nature Reserve in Merseyside, is an example.

## Managed coastal realignment

Managed realignment (also called managed natural retreat) means allowing the sea to flood areas that were previously protected. In time, salt marshes and mudflats form a zone that provides a natural defence to the 'new' shore, landwards of the abandoned area.

Managed realignment makes good sense where hinterland usage or development does not prevent it. It provides a way to compensate for the impacts of sea level rise and encourages natural environmental processes. The high and growing cost of schemes to prevent erosion and flooding mean that alternative approaches need to be pioneered. This approach was developed as an option only in the mid 1990s, so coastal managers are still learning about it.

The success or otherwise of particular schemes depends on a complex set of variables:

- estuary size, shape and location
- degree of wave exposure needed to reduce transport of silt from flooded fields
- tidal regime
- current land use and value, as this influences compensation and costs claims
- the land should, once designated, not be farmed using **agrochemicals**
- public perception is that this type of solution is 'giving in to the sea'
- the quality of the chosen site — i.e. its height above sea level, whether it is adjacent to an existing marsh area that can act as a seed bank, and has suitable hydrological conditions that can be managed to prevent scouring and allow sufficient water to facilitate sedimentation — otherwise there is potential for failure
- well researched site 'science', with accurate details and understanding of hydrological, sedimentological and ecological processes
- good site management to ensure that the breaches of the existing walls are correctly sited, with relic creek networks present as new man-made channels

**Exam tip**

Note that for WJEC and Eduqas you have to study *one* management strategy to manage the impacts of human activity on coastal processes and landforms.

**Exam tip**

Try to look at example areas outside the UK too, such as the Florida Coastal Management Program, which manages the impacts of development on mangroves, seagrass beds and coral, or the Great Barrier Reef Management Plans, which give a detailed account of zoning.

The anticipated consequences of such a revolutionary approach are many and generally positive:

- It is more sustainable, producing habitats of high ecological value, such as new salt marshes and mudflats.
- It encourages 'natural' protection, for example through mudflats and salt marshes, to counter **coastal squeeze**.
- It provides a possible reduction in the costs of protection, so much depends on the value of the land (below £18,000 per hectare, with limited coastal settlement and economic activity preferable).

However, these schemes are relatively new, so they are controversial and their full impacts are not yet known. Examples include the Essex coast (e.g. Abbotts Hall Farm) and Pagham Harbour (southern England), the North Sea coast (Netherlands), and the coasts of Louisiana and Florida (USA). Table 8 assesses the benefits and potential problems of this management option.

Table 8 The benefits and potential problems of managed realignment

| Benefits | Potential problems |
|---|---|
| <ul><li>Increased intertidal width and wave attenuation capacity</li><li>Increased conservation potential as there are additional habitats, many of high quality and ecofriendly</li><li>Retained naturalness of estuary zones</li><li>Increased protection against sea level rise and the avoidance of coastal squeeze</li><li>Potentially a cheaper option as low-value land with opportunities for visitor income (e.g. birding at Cley in north Norfolk) of up to £40,000 net per year (after maintenance costs)</li></ul> | <ul><li>Danger of incorrectly managing and modifying tidal processes</li><li>Experimental, novel technique that has some potential for failure</li><li>Complexity of potential compensation issues with farmers — so may have high start-up costs</li><li>Uncertainty about hydrology and sediment movements and adaptation to rising sea levels; complex to manage and possibility of disaster</li><li>While it is an obvious solution and has much to commend it the public have poor perceptions of its value, i.e. 'giving up', 'giving in to the sea'</li><li>Can only be used in certain estuary sites — not a universal cure</li><li>May need to be part of a wider estuary scheme rather than a piecemeal, small-scale, isolated scheme</li><li>Environmental values can be over-estimated, for example the impact on shellfish industries is still uncertain</li></ul> |

# Negative impacts of human activity

Several types of human activity are damaging to the coastal environment. The coastal zone is a contested zone, with increasing numbers of people having different opinions on how it should be used, managed, engineered and valued. It is hardly surprising, therefore, that there is conflict between conservation of the environment and the need to develop the coast economically.

## Offshore dredging

Dredging for aggregates (sand and gravel) takes place at varying scales, from small-scale abstraction to industrial-scale, commercial dredging operations. In most developed countries dredging is licensed because it is a potentially destructive activity to both marine ecosystems and the wider environment. Dr Douglas Clark, who specialises in the impacts of marine dredging, has identified dredging 'as the greatest single threat to coastal ecosystems'. For this reason, in most EU countries and Japan, dredging is restricted to areas landward of the 20 m **isobaths**.

> **Knowledge check 24**
>
> Evaluate the extent to which managed realignment has positive impacts as a coastal management option.

Dredged aggregates are used for shoreline protection, where they form hardcore for offshore breakwaters and other defences as well as being used for beach nourishment. Dredging is also frequently used to improve shipping access to ports or marinas by providing deeper navigable channels for even larger ships. The UK is particularly well endowed with abundant offshore gravel and sand deposits, as these are relict deposits from the Pleistocene.

Subaqueous dredging has huge impacts on the seabed. For example, a survey of offshore dredging along the coast of France showed that 160 km² were affected, causing serious ecological damage. The habitats of fish, invertebrates and algae are adversely affected, and benthic (bottom-dwelling) fauna are removed in the process, so disrupting the marine food webs, which in turn impacts on birds and mammals. Many popular dredging areas are the spawning grounds of fish, such as Dogger Bank in the North Sea, thus resulting in a decline in fish stocks.

In selective dredging, any small-calibre material that is not required is released back into the water. The fine sediment settles again after a long period of time but, in the meantime, the resulting high levels of turbidity suffocate filter feeders such as mussels. As harbours are deepened, surplus mud and other sediments (spoil disposal dumping) are taken out to sea and offloaded on the continental shelf.

Poisons (toxins), such as heavy metals and hydrogen sulfide, may be released from seabed sediment into the water column when dredging occurs. Nutrients are also released, which can lead to algal blooms.

Other environmental effects relate to the deepening of inshore waters, increasing shore-face slopes and therefore allowing larger waves to break close inshore, causing increased damage to the coast. Hallsands in Devon is a useful illustration of these impacts.

## Erosion of sand dunes

Sand dunes are dynamic systems, constantly adjusting to variations in wind patterns and sediment supply (page 31). Their presence and stability are of fundamental importance to coastal protection, for example off the Dutch coast, but they are threatened by human activity.

Dunes are highly fragile environments, formed from uncompacted sediment and poorly bound by vegetation, which, coupled with their exploitation for a variety of activities, makes them vulnerable to damage from overuse and misuse.

Human impacts on sand dunes can be grouped into four categories (Table 9).

Table 9 Categories of human impacts on dunes

| Conversion | Removal | Utilisation | External |
|---|---|---|---|
| Urbanisation Golf courses Agriculture Forestry | Mining Development | Tourism Trampling Horse riding Sand yachting Off-road vehicles Water extraction Conservation Military training | Reduced sediment supply Sea defences Dune migration prevention |

1 **Conversion** involves changing the vegetation type (e.g. afforestation, agriculture or golf courses) or the nature of the dunes by urbanisation and development. The dunes cease to function in their natural manner, either because sediment cannot be blown inland or because the dunes are built on.

2 **Removal** is where the sand is removed for other uses, for example mined for glass making, or when dune profiles are altered to facilitate beach access or provide a sea view.

3 **Utilisation** involves using the current resources. All types of tourism use the dune as an **amenity** with often disastrous effects on 'dune health' if the area is not managed. However, even dune conservation can have a negative impact. For example, where blowouts are artificially created to encourage growth of marram grass, this can lead to escalating rates of destruction.

4 **External** impacts include those from beyond the dune environment. Dunes can only survive with an adequate supply of sediment — any activity that inhibits this, such as building coastal defences, induces net dune decay, as they lose their ability to build and migrate.

Dunes tend to have a low threshold of survival, with only low levels of interference required to induce negative changes.

When development takes place on sand dunes for housing or holiday accommodation, all the available sand is sealed below the tarmac, rendering the dunes immobile. There is a loss of naturalness of the remaining surrounding dunes, and habitat fragmentation and habitat loss also occur. Constant dune encroachment on developments triggers further demands for protection, especially from future flooding risks. The dunes suffer knock-on damage from holiday campsites, evening barbeques, general vandalism and overuse by trampling, for example at Island Beach, New Jersey, USA. Mechanised sand mining, as on the southern Brittany coast near Morbihan, can lead to dune degradation and erosion as removal takes place at a greater rate than renewal.

Experiments have shown that trampling at even a low level of, say, 150 'passes' (i.e. 75 return journeys) per month can lead to a 50% reduction in vegetation, creating bare ground and the development of blowouts (e.g. South Beach, Blackpool).

Water extraction can lower the water table, thus rendering the dunes more liable to wind erosion. Also, as water is extracted, salt water is drawn in — and saline intrusion can damage the dune slacks.

Management strategies can be developed to stabilise and reconstruct the coastal dunes, as well as to control usage, although the heavier the usage the more difficult management becomes. There are three approaches:

1 Heavily degraded areas need **complete reconstruction**.

2 On less severely degraded dunes **restoration and repair** of the seaward face by replanting and fencing can be carried out.

3 The cause of the problem can be tackled with holistic solutions.

The 7 km$^2$ Ainsdale sand dunes on the northwest coast of England attract large numbers of visitors. Walking in the dunes, driving quadbikes and 4×4 vehicles,

**Knowledge check 25**

Explain why sand dunes have a low threshold of survival against damaging actions.

and barbeque fires are all eroding the dunes and destroying the vegetation. The management strategy for the area includes a zoning system that recognises honeypot areas. In these areas intense management occurs, providing facilities such as car parking and boardwalks to prevent erosion. The dune area forming the National Nature Reserve has limited access for permit holders only. In other areas fences are used to protect the vegetation by limiting access, and areas are replanted.

At Kenfig, South Wales, the sediment supply was reduced by the building of a jetty at Port Talbot. This, combined with sand extraction from the dunes and nearshore dredging, reduced the sediment supply to the dunes, leading to a negative sediment budget. A **holistic management** plan (walkways, access permits and keeping out grazing livestock) was the answer.

## Strategies to manage the impacts of human activity on coastal processes and landforms

The choice of a suitable case study to illustrate this concept is wide. For this guide, new, more sustainable coastal management strategies have been selected, with a detailed study of the Shoreline Management Plan for Barton on Sea, one of the most problematic coastal stretches on the south coast of the UK. An equally relevant approach would be to take a topic such as coastal conflicts caused by human activities and look at a plan to manage them at a small scale, such as issues caused by tourism at the Golfe de Morbihan in Brittany, or conflicts at Hengistbury Head on the UK's south coast, or using a fieldwork-based study from the Welsh coast.

> **Exam tip**
>
> Try to use an existing case study you know well.

### New, more sustainable ways to manage the coast

In the past, coastal management tended to deal with a single issue at a time, for example sediment loss from a particular beach, or collapse from landslip on a cliff face. Each issue was dealt with piecemeal by an individual authority. Although the original issue was often resolved, unintended consequences frequently arose — for example, sediment starvation and subsequently increased erosion downdrift.

**Integrated coastal zone management (ICZM)** overcomes this piecemeal approach and is now the preferred strategy in most countries of the world, with the development of **holistic management** policies that encompass a wide range of methodologies. It is now recognised that the geographical context of a coastline includes cliffs, beaches, dunes, marshes and estuaries, i.e. the **nearshore** areas *and* river catchments draining into the coast, as well as offshore areas. Equally, human impacts on the coast are part of the coastal system. It is important to bring together a wide range of groups and individuals, as there is an increase in the number and variety of stakeholders in the coastal zone.

Increasingly, ICZM has to be integrated **internationally** (i.e. transboundary), for example between nations adjacent to the North Sea or the Mediterranean Sea, because of wider sediment flows. Agenda 21 commits coastal nations to the implementation of ICZM initiatives, and to the sustainable development of coastal areas and marine environments under their jurisdiction. With the growing threats of climate warming, and the subsequent accelerating rise in sea level, there is clearly some urgency in the situation.

Sustainable management of the coast is increasingly the primary aim, balancing environmental needs with economic viability and social needs of both coastal and non-coastal dwellers (Table 10). While this is an admirable aim it is often ill-defined, and not easy to achieve in what is a contested environment.

Table 10 The coastal sustainability quadrant

| Futurity | Public participation |
|---|---|
| Using coastal areas for the benefit of the present population, while maintaining their potential for future generations | Ensuring that public stakeholders and individuals have the information and opportunity to take part in the decision-making process about choice of strategy |
| **Environmental and ecofriendly** | **Equity and social justice** |
| Providing coastal defences that enhance the environment and work with the natural environment | Ensuring that as far as possible the needs of all groups are met when considering options for coastal development, including poor people who have little influence over the decisions |

**Knowledge check 26**

Explain why sustainable coastal management is so difficult to achieve.

## Case study

### One management strategy to manage the impacts of human activity on coastal processes and landforms

Integrated coastal zone management: the shoreline management plan (SMP) for Barton on Sea

Sea defences built in the past to halt coastal erosion are today impacting the processes, and therefore the landforms, in the area.

Figure 20 shows the three coastal behavioural units (cliffs in the case of Barton on Sea) for a 6–7 km stretch of the Hampshire coast at Barton on Sea in Christchurch Bay, for which there are three different proposed options resulting from differences in the nature of the coastal zone and the subsequent strategic aims. Table 11 summarises these.

The geology of Christchurch Bay is mainly permeable sedimentary tertiary sands and gravels. At Barton on Sea, the underlying clay is exposed. Mass movement in the form of rotational slumps is caused by these adjacent layers of permeable sands and impermeable clays. Groundwater flows towards the sea from 400 m inland, which increases cliff face erosion and instability.

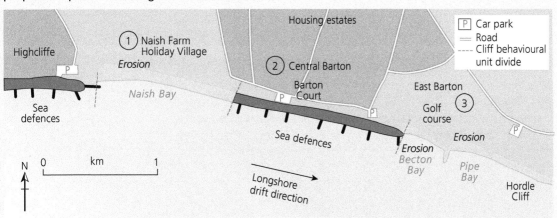

Figure 20 The coast at Barton on Sea

Table 11 Cliff behavioural units and issues: the situation by 2006

| Cliff behavioural units and issues | Management options and decisions |
|---|---|
| **1 West to Naish Farm Holiday Village and Highcliffe**<br>Area is a retreating and collapsing cliff, with good exposure of fossiliferous Barton Clay, hence its SSSI conservation status. Active cliff recession escalated from 0.4 m per year in the 1950s to 1.9 m per year in the 1970s, partly from the 'terminal scour' resulting from Highcliffe defences to the west. If no management is carried out, erosion may seriously affect the main Barton defences. | No major sea defences because of low-value cliff-top use (mobile caravans), and need for fresh fossil exposures and some sediment to feed Barton and Hurst Castle spit downdrift to the east.<br>The preferred SMP defence option for this zone is **'managed retreat'** in the central part of the zone while retaining the rock defences at its eastern and western boundaries at Barton and Highcliffe, respectively. Soft defence techniques, such as shingle recharge, will allow some cliff erosion to continue while keeping the beach front in its present location. |
| **2 Central Barton on Sea**<br>Barren, gravelly, defended cliffs, despite being part of SSSI, because of risk to property. An area of previously high erosion rate of >1 m per year now reduced by extensive coastal protection work since the 1930s, but still suffering periodic collapses of parts of the cliffs, as in 2001. | The main focus for coastal defences: no new developments at cliff edge (designated a green belt by the New Forest District Council, effectively a type of **red lining** to show on which side planning permission will be given/refused). New or redeveloped properties up to 400 m inland must have special soakaways to reduce groundwater build-up.<br>The preferred SMP option is to **'hold the existing defence line'**, justified by a cost–benefit ratio from the loss of economic value of properties inland — but needing major grant aid from Defra to fund a revolutionary inland drainage scheme, from 2007. |
| **3 East of Barton to Becton Bunny**<br>Still part of the SSSI. Sea defences at Barton, including rock groynes, trap longshore drift so there are still major retreats in the cliffs but losses only to the golf course. | The SMP option is to **'do nothing'** because of the low value of development inland, with rerouting of the coastal footpath. |

## Summary

- Positive impacts of human activity on coastal processes and landforms include management and conservation.
- Protection strategies operate at different scales, including coastal World Heritage Sites at a global scale, marine reserves at a national scale and SSSIs at a local scale.
- Negative impacts of human activity on coastal processes and landforms include offshore dredging and erosion of sand dunes.
- Dredging for aggregates has adverse impacts on the seabed, marine habitats and food webs. Poisons and nutrients may also be released. The deepening of inshore waters allows larger waves to break close inshore, damaging the coast.
- Sand dunes are important for coastal protection, but they are threatened by changes in vegetation type, urbanisation, development, recreational uses and removal of sand.

# Tectonic hazards

# Tectonic processes and hazards

## The structure of the Earth

The Earth's structure has been analysed by scientists studying patterns of shockwaves (caused by earthquakes). They have identified a number of layers, with different densities, chemical compositions and physical properties (Figure 1).

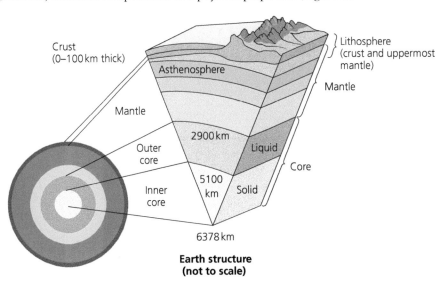

**Earth structure (not to scale)**

Figure 1 The Earth's structure

The Earth can be divided into three layers: the core, the mantle and the crust, based on their density and chemical composition.

- The **core** consists of iron and nickel and is about the size of the planet Mars. The outer core is in a semi-molten state but the inner core is solid. The temperature at the centre of the Earth (6378 km below the Earth's surface) is about 6200°C (even hotter than the surface of the Sun).
- The **mantle** is composed mainly of silicate rocks, rich in iron and magnesium. Apart from the solid top layer (known as the **asthenosphere**), the rocks in the mantle are in a semi-molten state. The mantle extends to a depth of 2900 km, where temperatures may exceed 5000°C. It is this high temperature that generates **convection currents**, which were identified as a mechanism driving plate movements.
- In relative terms, the **crust** is as thin as the skin of an apple is to its flesh. The crust is divided into:
  - **oceanic crust** (known as **sima** because it is composed predominantly of **si**lica and **ma**gnesium), a layer consisting mainly of basalt; it averages 6–10 km in thickness and at its deepest point has a temperature of 1200°C
  - **continental crust** (**sial**, so called as it is composed of **si**lica and **al**umina); it can be up to 70 km thick, and consists mainly of granite

Table 1 summarises the differences between the two types of crust.

Table 1 Differences between the two types of crust

|  | Oceanic crust | Continental crust |
| --- | --- | --- |
| Maximum age | 180 million years | 3.5 billion years |
| Thickness (km) | 6–10 | 25–75 |
| Area of the Earth's surface | 60% | 40% |
| Density (g cm$^{-3}$) | 3.3 | 2.7 |
| Rock type | Basaltic | Granitic |

The crust is separated from the mantle by the **Moho discontinuity** (named after Mohorovičić, the Croatian scientist who first discovered it).

The crust and the rigid top layers of the mantle are collectively known as the **lithosphere**.

## The mechanics of plate tectonics

The theory of plate tectonics states that the Earth's surface is made up of rigid lithospheric plates (seven major plates, seven smaller, minor plates, and many more even smaller ones known as micro-plates). There are some areas where the pattern of plate boundaries is so complex, such as Iran and Indonesia, that they appear rather similar to the smashed shells of hard-boiled eggs. As can be seen in Figure 2, some plates contain largely continental crust (e.g. the Eurasian plate), some are composed of continental and oceanic crust, and others contain only oceanic crust (e.g. the Nazca plate).

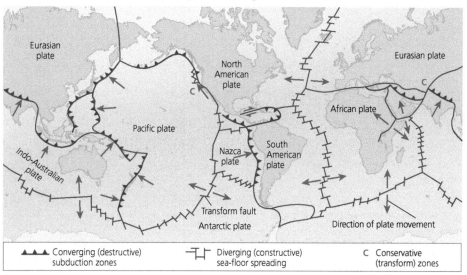

Figure 2 The Earth's main tectonic plates

The original idea was that the rising limbs of convection cells (at the spreading ridge) move heat from the Earth's core towards the surface, spreading out either side of the ridge and carrying the plates with them. The plates 'float' on a lubricated layer between the upper mantle and the lithosphere — the asthenosphere. This lubricated layer allows the solid lithosphere to move over the upper mantle (Figure 3).

**Knowledge check 1**

In which layer of the Earth's structure is the asthenosphere found?

**Exam tip**

Tectonic plates are composed of the lithosphere and not just the Earth's crust.

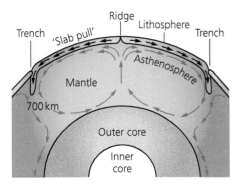

**Figure 3** The role of convection currents and slab pull

## Modern thinking on plate movement

Modern imaging techniques (tomography) have been unable to identify convection cells in the mantle that are sufficiently large to drive plate movement, so the idea of the asthenosphere as a 'conveyor' of plates has been modified. It has also been discovered that the injection of fresh magma associated with sea floor spreading at the ocean ridges does not **push** the plates apart, instead it is more of a passive process — a case of filling a gap rather than forcibly injecting material into the lithosphere.

Molten material wells up at diverging (constructive) plate margins because of thinning of the lithosphere, and the consequent decrease in pressure results in partial melting of the upper mantle. As the lithosphere is heated, it rises and becomes elevated above the surrounding sea floor to form an ocean ridge. This elevation produces a slope down and away from the ridge. Fresh rock formed at the spreading centres is relatively hot, less dense and more buoyant than rock further away from the diverging (constructive) margin, which becomes increasingly older, cooler and denser.

**Gravity** acts on this older, denser lithosphere, causing it to slide away from the spreading ridge. As a result, the lithosphere is thinned at the ridge, creating yet more partial melting and upwelling of magma. This process, known as **ridge push**, was initially identified as the key driver of movement but it is now considered to be a passive process — it is the **gravitational sliding** that is now regarded to be the active force driving plate movement. The density differences across the plates are of key importance. **Slab pull** occurs at subduction zones where the colder, denser portions of the plates sink into the mantle, and this **pulls** the remainder of the plates along. Therefore, slab pull is the key mechanism for plate movement. Evidence from tomography (seismic scans) supports this theory because cold, dense slabs of plate have been identified deep in the Earth's mantle.

The development of plate tectonics as a theory evolved over many years from a concept to a credible mechanism, but it remains just a theory because the cost of drilling down for proof is prohibitively expensive and impractical. The mechanism by which tectonic plates move is highly complex and remains a subject of debate.

**Exam tip**

When answering questions on plate tectonics do not include the story of continental drift in your answer.

## Plate movements

Plates move slowly and irregularly in relation to each other, typically at rates of 4 cm per year. Three types of movement are recognised:

1 In some locations plates move away from each other, i.e. they **diverge** at a **diverging (constructive) plate margin**, for example the East African Rift Valley.

2 In other locations plates move towards each other, i.e. they **converge** at a **converging (destructive) plate margin**, for example off the coast of South America.

3 In a few places plates move past each other, either in opposite directions, or in the same direction at different speeds, i.e. a **transform** movement, at what is called a **conservative (transform) plate margin**, for example the San Andreas fault, California.

## Processes operating at plate margins

Table 2 summarises the main settings, processes, hazards and landforms associated with plate margins. These settings are fundamental in explaining the spatial distribution and occurrence of nearly all tectonic hazards and landforms (see page 67 for exceptions).

**Table 2** Tectonic settings

| Tectonic setting | Motion (processes) | Hazards | Example | Landforms |
|---|---|---|---|---|
| Diverging (constructive) plate margins | Two oceanic plates moving apart | Basaltic volcanoes and minor, shallow earthquakes | Mid-Atlantic ridge (Iceland), mostly submerged | Lava plateaux, ocean ridge features |
| | Two continental plates moving apart | Basaltic spatter cones and minor earthquakes | Mt Nyiragongo (DRC) in the East African Rift Valley | Rift valley landscapes |
| Converging (destructive) plate margins | Two oceanic plates in collision | Island arcs, explosive andesitic eruptions and earthquakes | Soufrière Hills on Montserrat; Aleutian Islands | Island arcs of volcanoes |
| | Two continental plates in collision | Major, shallow earthquakes, long thrust faults | Himalayan orogenic belt collision zone | Compressional mountain belts |
| | Oceanic and continental plates in collision | Explosive andesitic eruptions and major earthquakes | Andes mountain chain and volcanoes | Complex mountain landscapes with fold mountains and volcanoes |
| Conservative (transform) plate margins | Plates sliding past one another | Major shallow earthquakes, no volcanic activity | San Andreas fault, California; North Anatolian fault, Haiti | Strike-slip faulted landscapes |
| Hotspots | Oceanic | Basaltic shield volcanoes and minor earthquakes | Hawaiian island chains; Galapagos Islands | Volcanic landscapes |
| | Continental | Colossal rhyolitic mega-eruptions | Yellowstone 'supervolcano', Wyoming | 'Roots' of super volcanoes |

**Exam tip**

Table 2 is a useful summary. Make sure you learn it.

**Knowledge check 2**

Which tectonic settings have (a) the most violent earthquakes, (b) the most explosive volcanic eruptions?

# Diverging (constructive) plate margins

## Two plates of oceanic crust

The movement apart of the plates is due to the divergence driven by **slab pull**, which brings magma from the asthenosphere to the surface. The pressure from the margins leads to a doming up of the Earth's surface and the formation of a ridge, such as the mid-Atlantic ridge (Figure 4). This ridge and rift system extends along the mid Atlantic for about 10,000 km. It was created about 60 million years ago when Greenland (on the North American plate) and northwest Scotland (on the Eurasian plate) separated to form the Atlantic Ocean. The **average** rate of movement is 0.025 m per year. There is a series of underwater volcanoes along the margin, which occasionally form a volcanic island. Iceland is one such volcanic island, much of which formed from a lava plateau about 200 m above sea level, as basic lava poured out through numerous tensional faults (fissures) formed by a hotspot plume. Subsidence of sections of crust between fault lines formed the rift valley, clearly visible at Thingvellir, and there are also active volcanoes such as Hekla and Grimsvotn linked to individual vents from the hotspots.

Most of the earthquake activity at a diverging (constructive) margin is shallow, low magnitude and high frequency, often along transform faults as the mid-ocean ridge is offset. In June 2000 a significant earthquake, MM scale 6.5, occurred on the south coast of Iceland.

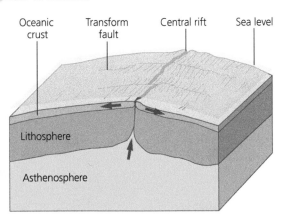

**Figure 4** Cross-section of the mid-Atlantic ridge

## Two plates of continental crust

Where two continental plates diverge the brittle crust fractures, forming parallel faults. Between the faults the crust slowly subsides, forming a rift valley. The East African Rift Valley is an example of a diverging (constructive) margin in an area of continental crust. Eastern Africa is moving in a northeasterly direction, diverging from the main African plate, which is 'heading' north. The rift valley, which consists of two broadly parallel branches, extends for 4000 km from Mozambique to the Red Sea. Inward-facing fault-line scarps (eroded fault scarps) reach heights of more than 600 m above the valley floor.

**Knowledge check 3**

Explain the difference between a fault scarp and a fault-line scarp.

Figure 5 shows the following features:

■ Linear mountain ranges (ridges) form as a result of the buoyancy of hot, low-density margins, which forces the crust to bulge upwards along the plate margins.

■ A central rift in the ridge forms because of subsidence between normal faults to form a rift valley.

■ Chains of lakes (e.g. Lake Tanganyika) form in the basins as the rift opens up.

■ Fissure eruptions occur from a series of basaltic lava flows.

■ Some large volcanoes (e.g. Kilimanjaro) form where the crust has thinned by tension and rising magma is extruded through the weaknesses.

■ Numerous small basaltic cinder cones form on the rift valley floor, often made of lava and some ash (composite).

■ Minor igneous intrusions occur, flowing up through the faults and fissures to form dykes.

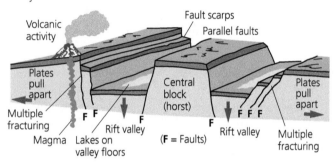

**Figure 5** Key features of the rift valley landscape

Over time the rift valley is reshaped, for example by waterfalls cascading over the lava plateau and by present-day weathering and mass movement on the fault scarp to create fault-line scarps.

## Converging (destructive) plate margins

Converging (destructive) plate margins occur when two plates converge due to slab pull.

### Ocean crust to ocean crust

When two oceanic plates converge, subduction occurs, as one plate is likely to be slightly older, colder or denser than the other. This plate is **subducted**, heated and eventually melts under pressure at around 100 km below the surface. The melted material rises up through any lines of weakness towards the surface. Extrusive volcanic activity results in the formation of a chain of volcanic islands above the subduction zone, known as an **island arc**. As Figure 6 shows, the Mariana Islands have been formed in this way through the convergence of the Pacific plate and the Philippines plate, with the Pacific plate being subducted to form the deep Marianas Trench. Earthquakes of high magnitude are focused along the subducted plate (the **Benioff zone**).

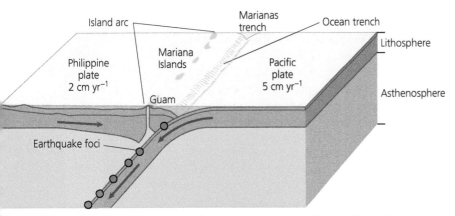

**Figure 6** Cross-section of an oceanic/oceanic converging (destructive) plate margin

## Oceanic to continental

Oceanic crust is denser than continental crust, so where these two types of crust converge, the more dense crust is subducted down into the asthenosphere by slab pull. Again, an ocean trench is formed on the sea floor at the point of **subduction**. The continental crust, because it is lighter and more buoyant, is not subducted but is uplifted and buckled, faulted and folded to form a range of mountains (Figure 7). Rising magma again breaks through any lines of weakness to form volcanoes, with infrequent but violent eruptions, or may solidify beneath the surface, forming intrusive igneous rocks such as granite batholiths, which can be subsequently exposed by numerous years of erosion.

An example is found in South America, where the oceanic Nazca plate is moving east at approximately 12 cm per year and is converging with, and subducting beneath, the continental South American plate, which is moving west at 1 cm per year. The Andes, a chain of fold mountains, rises nearly 7000 m above sea level, interspersed with many active volcanoes, such as Cotopaxi in Ecuador. The Peru–Chile trench, which reaches depths of 8000 m, occurs at the point of subduction. Earthquakes (e.g. in northern Peru in 1970 and in Ecuador in 2016) are frequent and often of huge magnitude (up to MM 9) and occur at a range of depths along the Benioff zone.

**Exam tip**

Learn easy-to-sketch annotated diagrams of all the major types of plate movement.

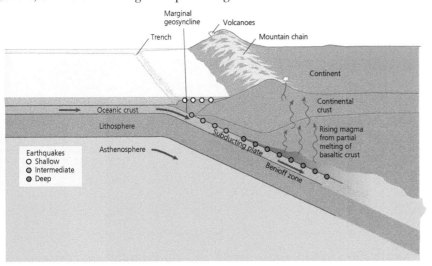

**Figure 7** Cross-section of an oceanic/continental converging (destructive) plate margin

*Continental to continental: collision margins*

Where two plates of continental crust converge is known as a **collision margin**. As the plates are both buoyant and composed of lower-density granite material, no subduction occurs. However, former ocean sediments trapped between the two converging plates are heaved upward, under intense compression (formation of **thrust faults** and nappes), resulting in the formation of major complex mountain belts. Usually no volcanic activity is found at this type of margin as no crust is being destroyed by subduction and no new crust is being created by rising magma. However, earthquakes do occur. Some are of deeper focus and therefore have less surface impact, but shallow, highly hazardous earthquakes also occur, often in populated foothill areas (e.g. Nepal).

A good example of a collision zone is the Himalayas. The Indo-Australian plate is moving northwards at a rate of about 5–6 cm per year, thus colliding with the Eurasian plate. Prior to their collision, the two continental landmasses were separated by the remnants of the Tethys Sea, which originated when Pangaea broke up around 300 million years ago.

As the two plates collided, the Himalayas were formed (orogenesis). These mountain belts are geologically complex as the intense compression has caused not only extreme folds (nappes), but also thrust faults and accretionary wedging on uplift. The mountain belt rises to heights of 9000 m and includes Mt Everest. The huge thickness of sediment has forced the crust downwards (isostatic depression) and the roots of the mountain belt are found deep in the Earth's interior. The collision movement causes great stresses, which are released by periodic earthquakes, such as the 2015 Nepal earthquake.

**Knowledge check 4**

Define the term 'orogenesis'.

## Conservative (transform) plate margins

A conservative (transform) margin is found where two plates move laterally past each other — this is known as a **transform movement**. As at collision margins there is no volcanic activity here because no crust is being destroyed by subduction and no new crust is being created by rising magma. However, shallow earthquakes of varying frequency and magnitude do occur.

High-frequency, low-magnitude earthquakes occur when pressure along the margin is relatively easily released, usually up to ten every day. Occasional major events take place after a significant build-up of pressure, typically when high levels of friction restrict movement along the original fault lines (e.g. the 2010 Haiti earthquake.

The best-known example of a conservative (transform) margin occurs in California at the San Andreas fault, where the Pacific plate and North American plate meet. The Pacific plate is moving northwestwards at a rate of 6 cm per year, while the North American plate, although moving in the same general direction, is only moving at about 1 cm per year. While earth tremors are very common, a 'Big One', such as the San Francisco earthquakes in 1906 and 1989, occurs only rarely.

Figure 8 summarises the main features found at a conservative (transform) margin, which are largely associated with erosion along the fault line.

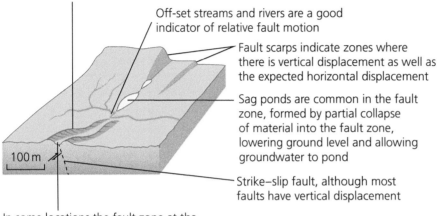

Compression ridges form where the fault has a turn or kink and the ground heaves upwards

Off-set streams and rivers are a good indicator of relative fault motion

Fault scarps indicate zones where there is vertical displacement as well as the expected horizontal displacement

Sag ponds are common in the fault zone, formed by partial collapse of material into the fault zone, lowering ground level and allowing groundwater to pond

Strike–slip fault, although most faults have vertical displacement

100 m

In some locations the fault zone at the surface forms a depression. The area is more easily eroded as the material is broken up into fault gouge

**Figure 8** The main features of a conservative (transform) margin

## Hotspots

Hotspots are small areas of the crust with an unusually high heat flow, and are found away from plate boundaries.

**Oceanic** hotspots occur where plumes of magma are rising from the asthenosphere. If the crust is particularly thin or weak, magma may escape onto the surface as a volcanic eruption. Lava may build up over time until it is above the present-day sea level, giving rise to volcanic islands.

The Hawaiian islands are a chain of volcanic islands (Figure 9) lying over a stable hotspot. The Pacific plate has been moving over the hotspot for about 70 million years, forming a succession of volcanic islands. As the plate has moved, so the volcanoes have been carried away from the hotspot in a northwesterly direction, forming a chain of extinct underwater volcanoes, called **seamounts**, extending all the way towards the Aleutian Islands. Currently a new volcano, called Loihi, is erupting 35 km southeast of Big Island (Hawaii). It is only 3000 m tall and has risen only to 2000 m below sea level to date — it is expected to reach the sea surface in 10,000–100,000 years' time. Big Island (Hawaii) volcanoes are extremely active — with frequent effusive eruptions from Kilauea. The high peaks Mauna Kea and Mauna Loa are actually higher than Mt Everest, but they start from well below sea level.

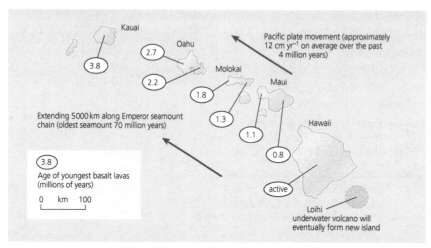

**Figure 9** The Hawaiian hotspot

An example of a **continental hotspot** is found beneath Yellowstone National Park. The key feature is the probable explosive eruption of a rhyolitic super volcano, at the maximum of the VEI scale (page 71). Geological records suggest that the Yellowstone event occurred 2.1 million years ago, ejecting 6000 times more gas and molten rock into the atmosphere than did the Mt St Helens eruption.

## Global distribution of tectonic hazards

Primary tectonic hazards include earthquakes and volcanoes.

### Earthquakes

Figure 10 shows the distribution of 30,000 earthquakes recorded over the last decade. The figure shows that the main zones of earthquakes are *not* randomly distributed but closely follow the plate boundaries.

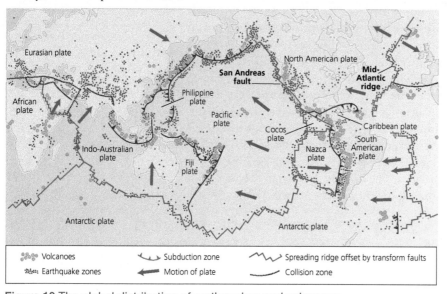

**Figure 10** The global distribution of earthquakes and volcanoes

**Exam tip**

Reread Table 2 (Tectonic settings) on page 60 to piece together the key facts regarding the distribution of earthquakes and volcanoes, which are summarised here.

**Exam tip**

Be able to name an example of a tectonic event that has occurred at each of the different types of plate margin.

The zones of earthquakes or **seismic** activity can be divided into four plate settings.

1　**Diverging (constructive)** margins along the ocean ridges. Earthquakes in this zone are mainly shallow, and result from tensional transform faults in the crust and from shaking during volcanic activity. Along the oceanic ridges many earthquakes are submarine and pose little risk to people.

2　**Converging (destructive)** margins, where oceanic crust is being subducted into the mantle beneath a continental plate, or where two oceanic plates collide in island arc zones. These areas are subject to frequent earthquakes, including high-magnitude ones, and represent areas of major hazards. Tsunamis are most commonly generated by these earthquakes (e.g. the 2004 Indian Ocean tsunami).

3　**Converging (destructive)** margins, where continental crust is colliding to produce fold mountain belts, e.g. the Alpine–Himalayan chain. Shallow earthquakes occur in a relatively broad zone, resulting in a high hazard risk (e.g. the 2003 Bam earthquake in Iran), with the occasional deep-seated earthquake.

4　Areas of lateral (**transform**) crust movement in the continental regions produce mainly shallow earthquakes of high magnitude, such as at the conservative (transform) margin of the San Andreas fault system in California.

Additionally, **intra-plate** earthquakes can occur — some 15% of all earthquakes occur in relatively stable continental crust, away from plate boundaries. These earthquakes are caused by stresses created in crustal rocks, usually by movement along ancient fault lines (e.g. New Madrid in Missouri, USA, 1811/1812 and Tangshen in China, 1976, which resulted in 240,000 deaths), in a process known as **isostatic recoil**. These intra-plate earthquakes are more dangerous because they are extremely unpredictable.

**Quasi-natural earthquakes** are those generated by human activity. It is thought that a 1993 earthquake in Killari, India, possibly resulted from the weight of water in a newly built reservoir behind a dam causing increased water pressure in the rock pores, which lubricated a fault line.

A recent controversial generator of earthquakes is **fracking** for unconventional supplies of oil and gas, which has led to numerous earthquakes in Oklahoma, USA, and also in north Lancashire, where fracking exploration was halted in 2015 as a result. Fracking was suspended in the UK in 2019.

## Volcanoes

The chemical composition of the lava depends on the geological situation in which it has formed. Basaltic (basic) lavas are formed by the melting of oceanic crust, whereas rhyolitic (acidic) lavas with a high silica content are formed from the melting of continental crust. Between these extremes are several groups of intermediate magmas, such as andesitic magma.

**Exam tip**

Always support your arguments with brief, locational fact files of examples.

**Exam tip**

Research how fracking can result in earthquakes.

**Knowledge check 5**

Explain the difference between magma and lava.

The world's active volcanoes are found in three tectonic situations (Figure 10).

1  **Diverging (constructive) margins (rift volcanoes)**. Most of the magma that reaches the Earth's surface (around 75% in quantity) is extruded along these boundaries. This mainly occurs at **mid-ocean ridges**, where melting of the upper mantle produces basaltic magma. The eruptions tend to be non-violent (VEI 1–2, page 71) and, as most occur on the sea floor, they do not represent a major hazard to people except where portions of the ocean ridge cross inhabited islands, such as Iceland. Fissure eruptions producing lava plateaux also occur widely. **Continental** diverging (constructive) margins, such as the East African Rift Valley system, also have active volcanoes, with a wide range of magma types depending on the local geological conditions through which the magma passes before reaching the surface.

2  **Converging (destructive) margins (subduction volcanoes)**. Some 80% of the world's active volcanoes occur along converging boundaries. As the oceanic plate is subducted into the mantle and melts under pressure, basic magma rises upwards and mixes with the continental crust to produce largely intermediate magma with a higher silica content than at the ocean ridges. These andesitic or, in some cases, more acidic, rhyolitic magmas can cause violent volcanic activity.

3  **Hotspots** — see page 65. Examples of active hotspots include the Hawaiian Islands, the Galapagos and the Azores. Eruptions are usually effusive with huge quantities of low-viscosity basaltic magmas, and are therefore less hazardous for people even in populated areas, although they can cause considerable damage to infrastructure and property.

## Physical hazard profiles and their impacts

Figure 11 compares the three major tectonic hazards in terms of their physical profiles. This is a qualitative technique that can be used to visually compare major hazard types, but also to look at a range of earthquakes or a series of volcanic eruptions.

**Exam tip**

Create a hazard profile diagram for the examples of different tectonic events you have studied.

**(a) Earthquake**

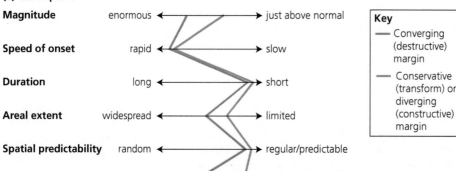

Figure 11 Tectonic hazard profiles: (a) earthquake, (b) volcano, (c) tsunami — a secondary hazard

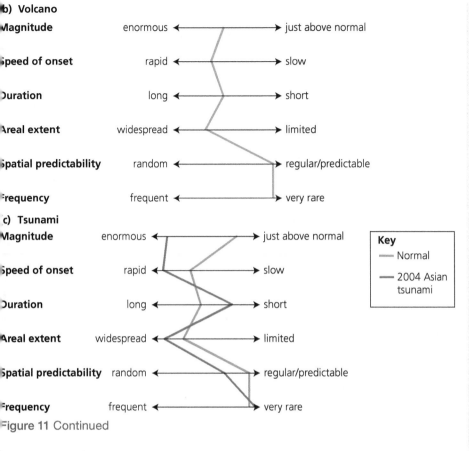

**b) Volcano**

| | | |
|---|---|---|
| Magnitude | enormous ← | → just above normal |
| Speed of onset | rapid ← | → slow |
| Duration | long ← | → short |
| Areal extent | widespread ← | → limited |
| Spatial predictability | random ← | → regular/predictable |
| Frequency | frequent ← | → very rare |

**c) Tsunami**

| | | |
|---|---|---|
| Magnitude | enormous ← | → just above normal |
| Speed of onset | rapid ← | → slow |
| Duration | long ← | → short |
| Areal extent | widespread ← | → limited |
| Spatial predictability | random ← | → regular/predictable |
| Frequency | frequent ← | → very rare |

**Key**
— Normal
— 2004 Asian tsunami

Figure 11 Continued

## Magnitude of earthquakes

**Magnitude** is considered to be the most important influence on the severity of impact of a tectonic hazard event. Magnitude is a quantifiable variable, especially for earthquakes. It can be defined as the size or physical force of a hazard event.

Earthquake magnitude is now measured by the **logarithmic moment magnitude (MM) scale**, a modification of the earlier **Richter scale**. The Richter scale was based upon the amplitude of lines made on a seismogram, using the largest wave amplitude recorded. So the bigger the earthquake, the greater the earth shaking. For example, a 1-unit increase on the scale represents a ten times larger amplitude, i.e. it is a **logarithmic scale**. The MM scale is based on a number of parameters of an earthquake event, including the area of fault rupture and the amount of fault movement involved, which determines the amount of energy released. Results are similar to the Richter scale, which is still widely used.

The **Mercalli scale** is also used to measure earthquakes. It is a descriptive scale that measures the amount of damage caused by surface shaking of particular earthquakes (Table 3). Table 4 shows the relationship between earthquake magnitude and number of resulting deaths in a sample decade.

**Exam tip**

Use the MM scale, now widely used, as it is an improved version of the Richter scale.

**Exam tip**

Try to learn simple tables of dates, locations, magnitudes and impacts of tectonic events.

Table 3 Abridged modified Mercalli intensity scale

| Average peak velocity (cm s⁻¹) | Intensity value and description |
|---|---|
| 1–2 | **I** Not felt, except by very few under exceptionally favourable circumstances.<br>**II** Felt only by a few persons at rest, especially on upper floors of buildings. Delicately suspended objects may swing. |
| 2–5 | **III** Felt quite noticeably indoors, especially on upper floors of buildings, but many people do not recognise it as an earthquake. Standing automobiles may rock slightly. Vibration like passing truck. Duration estimated. |
| 5–8 | **IV** During day felt indoors by many, outdoors by few. At night, some awakened. Dishes, windows, doors disturbed; walls make creaking sound. Sensation like heavy truck striking building. Standing automobiles rocked noticeably. |
| 8–12 | **V** Felt by nearly everyone, many awakened. Some dishes, windows etc. broken; cracked plaster in a few places; unstable objects overturned. Disturbance of trees, poles and other tall objects sometimes noticed. Pendulum clocks may stop.<br>**VI** Felt by all, many frightened and run outdoors. Some heavy furniture moved; a few instances of fallen plaster and damaged chimneys. Damage slight. |
| 20–30 | **VII** Everybody runs outdoors. Damage negligible in buildings of good design and construction; slight to moderate in well-built ordinary structure; considerable in poorly built or badly designed structures; some chimneys broken. Noticed by persons driving cars. |
| 45–55 | **VIII** Damage slight in specially designed structures; considerable in ordinary substantial buildings, with partial collapse; great in poorly built structures. Panel walls thrown out of frame structure. Fall of chimneys, factory stacks, columns, walls, monuments. Heavy furniture overturned. Sand and mud ejected in small amounts. Changes in well water. Persons driving cars disturbed. |
| > 60 | **IX** Damage considerable in specially designed structures; well-designed frame structures thrown out of plumb; damage great in substantial buildings, with partial collapse. Buildings shifted off foundations. Ground cracked conspicuously. Underground pipes broken.<br>**X** Some well-built wooden structures destroyed; most masonry and frame structures destroyed including foundations; ground badly cracked. Rails bent. Landslides considerable from riverbanks and steep slopes. Shifted sand and mud. Water splashed, slopped over banks.<br>**XI** Few, if any (masonry) structures remain standing. Bridges destroyed. Broad fissures in ground. Underground pipelines completely out of surface. Earth slumps and landslips in soft ground. Rails bent greatly.<br>**XII** Damage total. Waves seen on ground surface. Lines of sight and level distorted. Objects thrown into the air. |

Table 4 The relationship between earthquake magnitude and number of resulting deaths

| Date | Region | Magnitude | Deaths |
|---|---|---|---|
| 2004, 26 December | Sumatra and Indian Ocean | 9.1 | 227,898 |
| 2011, 11 March | Tohoku, Japan | 9.1 | 20,896 |
| 2008, 12 May | Eastern Sichuan, China | 7.9 | 87,587 |
| 2015, 25 April | Gorkha, Nepal | 7.8 | 8,964 |
| 2001, 26 January | Gujarat, northwest India | 7.7 | 20,085 |
| 2005, 8 October | Kashmir, east Pakistan/northwest India | 7.6 | 87,351 |
| 2018, 28 September | Sulawesi, Indonesia | 7.5 | 4,340 |
| 2010, 12 January | Port-au-Prince, Haiti | 7.0 | 230,000 |
| 2003, 26 December | Bam, southeast Iran | 6.6 | 26,271 |
| 2006, 26 May | Java, Indonesia | 6.3 | 5,782 |

## Magnitude of volcanic eruptions

All volcanoes are formed from molten material (magma) in the Earth's crust. There is no fully agreed scale for measuring the size of eruptions, but Newhall and Self (1982) drew up a semi-quantitative **volcanic explosivity index** (**VEI**), which can be related to the type of magma that influences the type of eruption. It combines the following into a basic 0–8 scale of increasing hazard:

- the total volume of ejected products
- the height of the eruption cloud
- the duration of the main eruptive phase
- several other items such as eruption rate

The results can be related to the type of volcanic eruption. For example, the 1991 eruption of Mt Pinatubo in the Philippines was a **Plinian type** of eruption with a plume of tephra ejected more than 30 km into the atmosphere. It was classified as a VEI 5–6. The VEI is, despite all the measurements, only a partly quantitative scale and it has several important limitations: for example, all types of ejected material are treated alike, and no account is taken of $SO_2$ emissions, which are needed to quantify the impact of eruptions on climate change. It is also difficult to apply as most of the very large eruptions (VEI 6–8) happened in geological time. Measuring the scale of volcanic eruptions is challenging because there are so many different types. Table 5 shows how volcanic eruptions can be classified using the VEI.

Note that the VEI is a logarithmic scale, because each step of the scale represents a ten-fold increase in material ejected.

**Table 5** The volcanic explosion index (VEI) scale

| Volcanic explosivity index (VEI) | Eruption rate (kg s$^{-1}$) | Volume of ejecta (m$^3$) | Eruption column height (km) | Duration of continuous blasts (h) | Troposphere/stratosphere injection | Qualitative description | Example |
|---|---|---|---|---|---|---|---|
| 0 Non-explosive | $10^2$–$10^3$ | <$10^4$ | 0.8–1.5 | <1 | Negligible/none | Effusive | Kilauea, erupts continuously |
| 1 Small | $10^3$–$10^4$ | $10^4$–$10^6$ | 1.5–2.8 | <1 | Minor/none | Gentle | Nyiragongo, 2002 |
| 2 Moderate | $10^4$–$10^5$ | $10^6$–$10^7$ | 2.8–5.5 | 1–6 | Moderate/none | Explosive | Galeras, Colombia, 1993 |
| 3 Moderate–large | $10^5$–$10^6$ | $10^7$–$10^8$ | 5.5–10.5 | 1–12 | Great/possible | Severe | Nevado del Ruiz, 1985 |
| 4 Large | $10^6$–$10^7$ | $10^8$–$10^9$ | 10.5–17.0 | 1–>12 | Great/definite | Violent | Mayon, 1895 Eyjafjallajökull, 2010 |
| 5 Very large | $10^7$–$10^8$ | $10^9$–$10^{10}$ | 17.0–28.0 | 6–>12 | Great/significant | Cataclysmic | Vesuvius, AD 79 Mt St Helens, 1980 |
| 6 Very large | $10^8$–$10^9$ | $10^{10}$–$10^{11}$ | 28.0–47.0 | >12 | Great/significant | Paroxysmal | Mt Pinatubo, 1991 |
| 7 Very large | >$10^9$ | $10^{11}$–$10^{12}$ | >47.0 | >12 | Great/significant | Colossal | Tambora, 1815 |
| 8 Very large | — | >$10^{12}$ | — | >12 | Great/significant | Terrific | Yellowstone, millions of years ago |

Magnitude is largely measurable (easier for earthquakes) and can clearly influence impact.

## Frequency

**Frequency** (i.e. how often an event occurs) is sometimes called the **recurrence** level, for example a 1 in 100 year event. There is an inverse relationship between frequency and magnitude, i.e. the larger the magnitude of the event, the less frequent its occurrence. The effect of frequency on severity of impact is difficult to gauge. Theoretically, areas that experience frequent tectonic events have both adaptation and mitigation measures in place, including extensive monitoring (useful for volcanoes), education and community awareness about what to do (useful for earthquakes or tsunami evacuation routes), and various technological strategies for shockproof building design (Tokyo, San Francisco) or protection (Japanese tsunami walls). It is well known that unexpected tectonic events, such as the 1993 Killari earthquake, can be particularly devastating. On the other hand, familiarity with a frequently erupting volcano, such as Mt Merapi in Indonesia, can breed contempt, as local people are so used to its eruptions that they do not always evacuate soon enough.

## Duration

**Duration** is the length of time for which the tectonic hazard exists. Often an initial earthquake event is followed by massive aftershocks (e.g. Christchurch, New Zealand, 2010 and central Italy, 2014) or a series of eruptions occurs. While individual earthquakes often last for only around 30 seconds, the damage can be extensive.

**Secondary hazards** often prolong the duration of impact and increase the damage, for example the Tohoku multi-disaster (earthquake, tsunami and nuclear accident). Secondary hazards associated with volcanic eruptions include **lahars** (e.g. Mt Pinatubo) or **jökulhlaups** (glacier bursts, see page 79), which are very damaging because of their spatial and temporal unpredictability. In November 1985, the melting of the ice cap and snow on the Nevado de Ruiz volcano released huge mudflows that overwhelmed Armero and the surrounding villages, killing 23,000 people. Locally, Himalayan earthquakes, such as Kashmir 2005 and Sichuan 2008, cause widespread landslides that disrupt rescue and recovery and add to the death toll.

## Areal extent

**Areal extent** is the size of the area covered by the tectonic hazard. This has a clear impact (Figure 12).

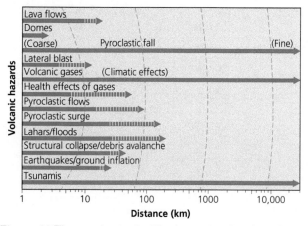

**Figure 12** The areal extent of the impacts of various features of a volcanic eruption

> ### Knowledge check 6
>
> Define the term 'aftershock' and explain its significance for the 2010 Christchurch earthquakes.

## Spatial concentration

**Spatial concentration** is the areal distribution of tectonic hazards over geographical space. It is controlled largely by type of plate boundary. In theory, permanent settlement is avoided in hazardous regions. However, often such locations present other opportunities, for example some volcanic soils are very fertile, so agricultural settlement occurs, such as on the flanks of Mt Merapi, Indonesia. Equally, spring water may be available, such as at Bam in Iran, the site of a severe earthquake. Active tectonic, especially volcanic, landscapes encourage tourism, as was seen in a recent unexpected Japanese eruption (Ontaki), where many of the 48 dead were hikers.

## Speed of onset

**Speed of onset** refers to how quickly the peak of the hazard event arrives, and can be a crucial factor. Earthquakes generally come with little warning. The speed of onset and the almost immediate shaking of the ground allows no time to issue warnings, and therefore has the potential for greater impact. This led to maximum destruction by the 1995 Kobe earthquake in Japan, but this was allied with other factors, such as timing and building type. The 2004 Indian Ocean tsunami illustrates the timing issue well, with little awareness of the hazard possible at Aceh, Indonesia, but, theoretically, warnings and therefore evacuation were possible everywhere else. This was in spite of a lack of a sophisticated warning system (subsequently built) for the Indian Ocean, unlike that which existed in the Pacific Ocean, located near Honolulu, Hawaii.

## Predictability of occurrence

The relationship between tectonic hazards and plate boundaries allows a level of prediction of location but not of when an event will occur. The random temporal distribution of both earthquakes and volcanoes can add to their potential impact. Accurate predictions can reduce the impact by allowing evacuations, while too many inaccurate predictions may increase the impact as people begin to ignore warnings. **Seismic gap theory** suggests that, over a period of time, all parts of a fault must attain the same average level of movement. This may be achieved by many minor events or is the result of a rarer but larger event. While gap theory can increase the possibility of predicting the 'Big One', in reality earthquakes are unpredictable. Volcanic eruptions can also be hard to predict precisely, even with close monitoring, hence discussions concerning the possible and long-awaited eruption of Vesuvius in the densely settled Bay of Naples.

**Knowledge check 7**

Explain seismic gap theory and its role in earthquake prediction.

## Summary

- The Earth's structure has a number of layers: the inner core, outer core, mantle and crust, each with different densities, chemical compositions and physical properties.
- The theory of plate tectonics states that the Earth's surface is made up of rigid plates 'floating' on the asthenosphere. Previously their movement was thought to be powered by convection currents but today gravitational sliding is considered to be the force driving plate movement, with slab pull being the key mechanism.

- Plates move in three ways: apart at a diverging (constructive) margin, towards each other at a converging (destructive) margin, or making a transform movement at a conservative (transform) margin.
- Hotspots are small areas of the crust with an unusually high heat flow, away from plate boundaries.
- The main zones of earthquakes closely follow the plate boundaries, but intra-plate earthquakes and quasi-natural earthquakes (generated by human activity) also occur.
- Active volcanoes are found along diverging (constructive) and converging (destructive) plate margins and at hotspots.
- The characteristics of the physical hazard profile that influence its impact include magnitude, predictability, frequency, duration, speed of onset and areal extent.

# Volcanoes, processes, hazards and their impacts

## Types of volcano

Volcanoes can be classified by their **shape** and the nature of the **vent** the magma is extruded through, as well as the nature of the **eruption**.

## Shape of the volcano and its vent

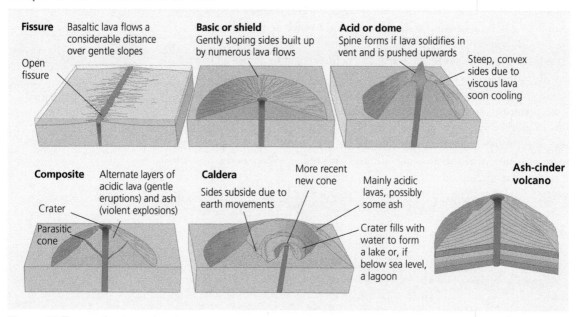

Figure 13 Types of volcano by shape

Figure 13 shows how volcanoes can be classified by shape. The shape is, of course, largely dependent on the material erupted, which itself can be linked back to the tectonic setting.

- **Fissure eruptions** result when lava is ejected through tensional, linear fissures, rather than a central vent, at diverging (constructive) margins. The 1973 Eldfell eruption in Heimaey, Iceland began with a fissure 2 km long, through which lava poured effusively.

- **Shield volcanoes**, such as Mauna Kea in Hawaii, are formed when basaltic lava pours out of a central vent in huge quantities. Because of the effusive nature of the basic fluid lava, it can spread over a wide area before solidifying. The result is a huge volcanic cone, but with gently sloping sides. The lava has low viscosity because of its low (<50%) silica content, and erupts at temperatures of about 1200°C. Mauna Loa in Hawaii has built up a cone that rises 9000 m from the sea floor, with a diameter of 120 km at its base, sloping at about 6° to the top. Shield volcanoes occur at both oceanic diverging (constructive) plate margins and hotspots.

- **Composite cones or strato-volcanoes** form from alternating layers of lava and ash, resulting from eruptions at converging (destructive) plate margins. The lava itself is typically acidic with more than 50% silica content. It has a temperature of about 800°C. This means it flows more slowly, creating cones with more steeply sloping sides. The ash is produced in a highly explosive eruptive phase, often after the vent has been blocked. Many of the world's well-known volcanoes, such as Mt Etna, Vesuvius and Popocatepetl, are composite volcanoes. Etna is a classic cone shape with slopes of 50° near its base, but slopes of only 30° at the summit, with a marked crater.

- **Acid or dome volcanoes** result when acid lava quickly solidifies on exposure to the air. These volcanoes frequently have parasitic cones, formed as the passage of the rising rhyolitic magma through the main vent is blocked. The cones are steep sided and convex in shape. In one extreme example (Mt Pelée, Martinique) the lava actually solidified as it came up the vent, producing a spine, rather than flowing down the sides.

- **Ash and cinder cones**, such as Paricutin in Mexico, are formed when ash and cinders build up in a cylindrical cone of relatively small size. Being composed of loose volcanic cinders means that they are highly permeable. A typical size is 800 m in height, with a bowl-shaped crater.

- **Calderas** occur when the build-up of gases becomes extreme. Huge explosions may clear the magma chamber beneath the volcano and remove the summit of the cone, or cauldron subsidence may occur. This causes the sides of the crater to collapse and subside, thus widening the opening to several kilometres in diameter. Frequently, enlarged craters or calderas have been flooded and later eruptions have formed smaller cones in the resulting lake, for example Wizard Island in Crater Lake, Oregon.

## Nature of the eruption

The nature of the eruption is also significant. Table 6 summarises the major categories of eruption, based on the degree of violence/explosivity, a consequence of pressure and the amount of gas in the magma.

Table 6 Major categories of eruption

|  | Category | Description |
|---|---|---|
| **Effusive** | Icelandic | Lava flows directly from a fissure |
|  | Hawaiian | Lava is emitted gently from a vent |
|  | Strombolian | Small but frequent eruptions occur |
|  | Vulcanian | More violent and less frequent eruptions |
|  | Vesuvian | Violent explosion after a long period of inactivity |
|  | Krakatoan | Exceptionally violent explosion |
|  | Peléan | Violent eruption of pyroclastic flows (nuées ardentes — see below) |
| **Explosive** | Plinian | Large amounts of lava and pyroclastic material are ejected |

> **Exam tip**
>
> Match the eruption types to the VEI scale. Always have a named example ready for each eruption type.

# Volcanic processes and the production of associated hazards

## Pyroclastic flows and surges

Pyroclastic flows have been responsible for most volcanic deaths to date. They are sometimes called **nuées ardentes** ('glowing clouds') and result from frothing of the molten magma in the volcano vent. Bubbles in the magma burst explosively to eject a lethal mixture of hot gases and pyroclastic material (volcanic fragments, ash, pumice and glass shapes). Pyroclastic bursts surge downhill because, as they contain a heavy load of rock fragment and dust, they are denser than the surrounding air. The clouds can be literally red hot (up to 1000°C). The greatest risks occur when the summit crater is blocked by viscous rhyolitic magma and blasts are directed laterally in Peléan type eruptions, in surges at $30\,\mathrm{m\,s^{-1}}$, close to the ground and reaching up to 30–40 km from the source.

There is little warning of these events; people exposed are killed immediately by severe external and internal burns combined with asphyxiation. The cloud that hit the town of St Pierre, Martinique (6 km from the centre of the VEI 6 eruption) in the Mont Pelée disaster of 1902 had a temperature of 700°C and travelled at $33\,\mathrm{m\,s^{-1}}$ down the River Blanche Valley. All but three of the inhabitants of St Pierre (around 30,000 in all) were killed. One was saved by being in the jail!

## Lava flows

While lava flows are spectacular, they pose more threat to property than to human life — for example, eruptions of Kilauea covered $78\,\mathrm{km^2}$, resulted in the destruction of much of the village of Kapilani, and destroyed nearly 200 houses. The lava flows most dangerous to human life come from fissure eruptions, not central vents, as highly fluid basalt magma can move down a hillside at $50\,\mathrm{km\,h^{-1}}$ and can spread a long way from the source. One deadly lava flow erupted from Nyiragongo volcano's flanks, draining the lava lake that had collected at the summit; it killed 72 people and devastated the town of Goma in the Democratic Republic of the Congo.

**Pahoehoe** lava is the most liquid of all lava, and tends to form a ropey, wrinkled surface. On steep slopes this low viscosity lava can move downhill at speeds approaching $50\,\mathrm{km\,h^{-1}}$.

**A'a** lava tends to form blocks, and moves more slowly downhill, leaving a rough, irregular surface.

The greatest lava-related disaster in historic times occurred in 1783 when huge quantities of lava poured out of the 24 km-long Laki fissure in Iceland. Although there were few direct deaths, the resultant famines from lack of crop growth killed more than 10,000 people — around 20% of Iceland's population.

## Airfall tephra (ash falls)

Tephra consists of all the fragmented material ejected by the volcano that subsequently falls to the ground. The large explosive eruptions of Mt St Helens (VEI 5) produced an estimated 6 km$^3$ volume of material, which covered a wide area of northwest USA. The particles ranged in size from so-called 'bombs' (>32 m in diameter) down to fine ash and dust (<4 mm in diameter). Coarser, heavier particles fell out of the sky close to the volcano vent. Occasionally tephra is sufficiently hot to spontaneously combust and start fires. Ash clouds can be blown many miles away from the original eruption by strong winds.

Large eruptions, such as Krakatoa (VEI 6) in 1883, which spread an aerosol cloud around the globe within 2 weeks, and Tambora, Indonesia in 1816 (VEI 7), which led to short-term global cooling of around 1–2 years, could be prolonged to a decadal scale by successive eruptions.

Although ash falls account for fewer than 5% of direct deaths associated with volcanic eruptions (usually respiratory problems), they can create a number of problems:

- Heavy falls of cinders and ash can blanket the landscape, contaminating farmland and poisoning livestock.
- Ash causes health issues such as skin abrasion and breathing problems (e.g. silicosis and chronic obstructive pulmonary disease (COPD)).
- The weight of ash can damage roofs.
- Ash washes into lakes and rivers to become a lahar source (page 78).
- Wet ash conducts electricity and can cause failure of electronic equipment.
- Fine ash can clog air filters and damage vehicles and aero engines.
- Ash can lead to vehicle accidents through poor visibility and slippery roads.

## Volcanic gases

Large amounts and a wide range of gases are released from explosive eruptions and from cooling lava. The complex gas mixture includes water vapour, hydrogen, carbon monoxide, carbon dioxide, hydrogen sulfate, sulfur dioxide, chlorine and hydrogen chloride, in variable amounts.

Carbon monoxide can cause death because of its toxic effects at very low concentrations, but most fatalities have been associated with carbon dioxide ($CO_2$) releases, because $CO_2$ is colourless and odourless. In Indonesia, as villagers were evacuating following the eruption of Mt Merapi, they walked into a dense pool of volcanically released $CO_2$ that had sunk (it is denser than air); 140 people were asphyxiated.

**Exam tip**

Research the 2010 Eyjafjallajökull eruption, which had a huge impact on the world economy because it brought flights to a standstill, with enormous global impacts. It is a unique case study. (See page 82.)

**Knowledge check 8**

Outline the differences between ash falls and pyroclastic flows.

The release of $CO_2$ from past volcanic activity can also create a highly unusual threat. In 1984 a cloud of gas, rich in $CO_2$, burst out of the volcanic crater of Lake Monoun, Cameroon, killing 37 people. Two years later, in 1986, a similar disaster occurred at the Lake Nyos crater, Cameroon, killing 1746 people and more than 8000 livestock. The outburst of gas created a fountain that reached 100 m above the lake, before the dense cloud flowed down two valleys to cover an area of more than 60 km².

These rare hazards are the result of unusually high levels of $CO_2$ in the volcanic lakes. The levels probably build up over a long period of time from $CO_2$-rich groundwater springs flowing into the submerged craters.

Sulfur dioxide can be a skin irritant, as well as being a greenhouse gas and thus having an impact on climate change. When it mixes with rainwater it can result in acid rain.

Figure 14 outlines the different hazards associated with volcanoes.

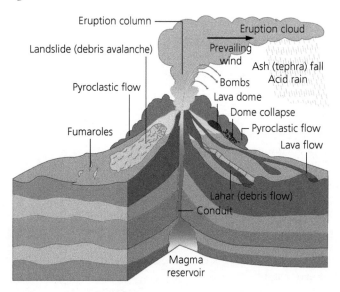

**Figure 14** Types of volcanic hazard

**Knowledge check 9**

Using examples, distinguish between primary, secondary and tertiary tectonic hazards.

## Secondary volcanic hazards

### Lahars

After pyroclastic flows, lahars present the greatest risk to human life. They can be defined as volcanic mudflows composed of largely silt-size sediments. Lahars consist of volcanic ash and rock up to at least 40% by weight, combined with the torrential rain that often accompanies volcanic eruptions. They create dense, viscous flows that can travel even faster than clear-water streams. They occur widely on steep volcanic flanks, especially in tropical humid or monsoon climates (**lahar** is an Indonesian term).

The degree of hazard varies greatly but generally the flows that contain the larger-size sediments are the more deadly. At Mt Pinatubo in the Philippines, lahars regularly transport and deposit tens of millions of cubic metres of sediment in a day and are a potential threat to the local population of over 100,000 people.

Lahars can be classified as **primary**, occurring directly during a volcanic eruption (usually **hot flows**), and **secondary**, which are triggered by high-intensity rainfall between eruptions, which reactivate old ash flows etc.

Some lahars are generated by rapid melting of snow and ice — a particular hazard in northern Andean volcanoes. The second deadliest disaster in recent times was caused by the 1985 eruption of the Nevado del Ruiz volcano, where a lahar killed over 20,000 people (page 82).

## *Volcanic landslides*

**Landslides** and debris avalanches are a common feature of volcano-related ground failure. They are particularly associated with eruptions of siliceous acidic dacitic magma of relatively high viscosity, with a high content of dissolved gas.

**Volcanic landslides** are gravity-driven slides of masses of rock and loose volcanic material. They can occur during an eruption, such as that at Mt St Helens, when the side of the volcano collapsed to form massive landslides and debris avalanches containing $2.7\,km^3$ of material. They can be triggered by heavy rainfall or, more commonly, earthquakes.

In 2018, the eruption of Anak Krakatau in Indonesia resulted in landslides and the collapse of the volcanic cone, which in turn triggered a tsunami resulting in the death of 437 people and displacing over 40,000.

**Ground deformation** of volcanic slopes by rising magma, which creates a bulge, can also trigger slope instability and landslides before an eruption, for example prior to the Mt St Helens events. Swarms of small earthquakes were followed by ground uplift and the formation of a huge bulge prior to the main eruption, and then the final trigger of a more major post-bulge earthquake.

Landslides can bury areas in debris many metres thick, destroying buildings and infrastructure. They may act as a source of material for the formation of lahars. It is possible that a landslide removing part of a cone can release pressure, triggering an eruption.

## *Jökulhlaups*

In most subglacial eruptions, the water produced from melting ice becomes trapped in a lake between the volcano and the overlying glacier. Eventually this water is released as a violent and potentially dangerous flood. As events of this type are so common in Iceland the Icelanders have coined the term jökulhlaup, which means glacial outburst. One of the most dramatic jökulhlaups to ever occur was a result of an eruption of Grimsvotn in 1996. Over one month, more than $3.2\,km^3$ of meltwater accumulated beneath the Vatnajökull ice cap. The subglacial lake suddenly burst out, with some of the water escaping beneath the ice cap and some spouting out through a side fissure. The lake drained in 40 hours, with an estimated peak discharge of $50,000\,m^3\,s^{-1}$, which is over 20 times the rate of flow at Niagara Falls. The resulting flood was temporarily the second biggest flow of water in the world (after the River

Amazon). It caused US$14 million of damage and left numerous icebergs scattered across Iceland's southern coastal plain. Generally, jökulhlaups rarely turn into disasters because they occur in remote unpopulated areas.

*Note:* **tsunamis** (page 86) can occur after catastrophic volcanic eruptions as well as earthquakes, but more rarely so — only 5% of tsunamis are generated by volcanic activity. For example, the structural failure of the volcanic island Krakatoa (VEI 6) in 1883 created a debris flow big enough to produce a tsunami.

# Impacts of volcanic hazards on people and the built environment

Volcanic hazards are volcanic events with the potential to cause harm, loss or detriment to humans, including the things humans value.

> hazard event × vulnerability of people = adverse consequences, harm or loss

The impact of volcanic hazards depends on a number of factors, including the **physical profile** of the volcanic event (page 69), the status of the volcano — whether it is extinct, dormant or active — and key factors such as population density, level of development, standard of governance, timing of eruption, and presence or absence of mitigation strategies. The impacts can be classified as both primary (direct results of the volcanic activity) and secondary (occur as a consequence of the primary impacts).

## Environmental impacts at local, regional and global scales

The **environmental** impacts of explosive volcanic eruptions are largely related to weather and climate, as shown in Table 7.

Table 7 The effects of large explosive volcanic eruptions on weather and climate

| Effect | Mechanism | Begins | Duration | Scale |
|---|---|---|---|---|
| Increased precipitation | H$_2$O given off in large quantities during eruption | During eruption | 1–4 days, i.e. period of eruption | Local |
| Reduction of diurnal cycle | Blockage of short-wave and emission of long-wave radiation | Immediately | 1–4 days | Local |
| Reduced tropical precipitation | Blockage of short-wave radiation, reduced evaporation | 1–3 months | 3–6 months | Regional |
| Summer cooling of northern hemisphere, tropics and subtropics | Blockage of short-wave radiation | 1–3 months | 1–2 years | Regional |
| Stratospheric warming | Stratospheric absorption of short-wave and long-wave radiation | 1–3 months | 1–2 years | Global |
| Winter warming of northern hemisphere continents | Stratospheric absorption of short-wave and long-wave radiation | 6 months | 1 or 2 winters | Regional |
| Global cooling | Blockage of short-wave radiation | Immediately | 1–3 years | Global |
| Global cooling from multiple eruptions | Blockage of short-wave radiation | Immediately | 10 years | Global |
| Ozone depletion, enhanced ultraviolet | Dilution, heterogeneous chemistry and aerosols | 1 day | 1–2 years | Global |

**Exam tip**

For A-level you must learn two contrasting examples of eruptions to demonstrate the variety of risks and impacts of volcanic activity. This includes impacts on the environment. For AS you need to learn one example, but do not have to study the environmental impacts.

**Knowledge check 10**

Define 'active', 'dormant' and 'extinct' volcanoes.

Volcanoes can also impact on the environment at the local scale by eruptions, lahars and landslides reshaping the landscape and, combined with lava flows and ash falls, covering the land and destroying the vegetation.

## Demographic, economic and social impacts of volcanic hazards on people and the built environment

Both primary and secondary effects of volcanic eruptions can have demographic, social and economic impacts (Table 8).

Table 8 The impacts of volcanic hazards

| | Demographic/social | Scale | Economic | Scale |
|---|---|---|---|---|
| **Primary impacts** | Over 91,000 killed by asphyxiation, thermal injuries and trauma | Regional | — | — |
| | Over 15,000 injured, 1900–2016 | Local | | |
| | Over 4.7 million affected, 1900–2016 | National | | |
| **Secondary impacts** | Over 15,000 a year made homeless | Local | Financial loss from loss of primary activities | Regional/national |
| | Income loss from destruction of farmland and businesses, plus the associated unemployment | Regional | Estimated cost of clean-up, repairs and rebuilding £50 million a year | National |
| | Health issues from lack of food and clean water; spread of disease | Regional | Estimated economic loss £2.7 billion, 1900–2016 | Global |

In general, volcanoes are comparatively minor hazards compared with other geo hazards, and in terms of their contributions to all natural disasters (Table 9).

Table 9 Average losses per year from volcanoes, earthquakes and all natural disasters 1975–2000

| Impact | Volcanoes | Earthquakes | All natural disasters |
|---|---|---|---|
| People dead | 1019 | 18,416 | 84,034 |
| People injured | 285 | 27,585 | 65,296 |
| People made homeless | 15,128 | 239,265 | 4,856,586 |
| People affected | 94,399 | 1,590,314 | 144,000,000 |
| Estimated damage (US$ billions) | 0.065 | 21.5 | 62.0 |

Figure 15 shows a map of the Nevado del Ruiz eruption, which devastated the local area and the overall economy of Colombia. Table 10 shows the damage table for this eruption.

**Exam tip**

Impacts should be illustrated by 'mini' case studies. These short fact files provide useful supporting evidence in an exam to add locational knowledge to your answer.

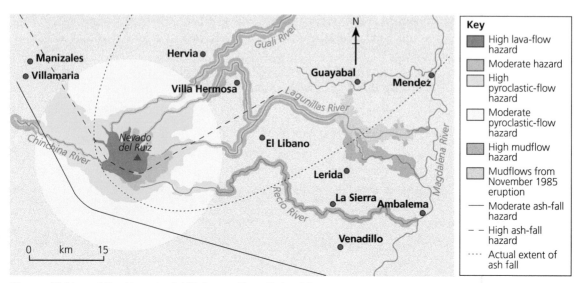

**Figure 15** Map of the Nevado del Ruiz eruption, Colombia

**Table 10** Damage table for the Nevado del Ruiz eruption

| Category of loss | Details |
|---|---|
| Deaths and injuries | Nearly 70% of the population of Armero killed (20,000 approx.) and a further 17% (5000) injured |
| Agricultural | 60% of the region's livestock, 30% of sorghum and rice crops, and 500,000 bags of coffee destroyed. Over 3400 ha of agricultural land lost from production |
| Communications | Virtually all roads, bridges, telephone lines and power supplies in the area destroyed. Whole region isolated |
| Industrial, commercial and civic buildings | 50 schools, two hospitals, 58 industrial plants, 343 commercial establishments and the National Coffee Research Centre badly damaged or destroyed |
| Housing | Most homes destroyed; 8000 people made homeless |
| Monetary | Cost to the economy estimated at US$7.7 billion, or 20% of the country's GNP for that year |

In contrast, Table 11 shows how the 2010 Eyjafjallajökull eruption in Iceland had a very different impact. A feature of the eruption was an ash plume 10 km high, which was blown southeast over a large part of Europe and the North Atlantic. The risk to aircraft resulted in hundreds of airports in more than 20 countries and a correspondingly large airspace in Europe being closed for a week.

**Table 11** Damage table for the Eyjafjallajökull eruption, Iceland

| Category of loss | Details |
|---|---|
| Deaths and injuries | 0 |
| Agricultural | Some farmland flooded or covered in ash; 20 farms destroyed; Kenyan farmers destroyed 3000 tonnes of flowers as they were unable to export them |
| Communications | Roads damaged by floodwaters. Closure of large parts of European airspace caused extensive air travel disruption |
| Industrial, commercial and civic buildings | Limited damage as ash cloud moved away from Reykjavik. Global air industry lost £130 million a day |
| Housing | 800 evacuated, but nobody made homeless |
| Monetary | $100 million economic loss to Iceland. $5 billion economic cost to the European economy |

**Exam tip**

Always apply your case studies to the specific question — avoid narrative description.

## Summary

- Volcanoes can be classified by shape, type of vent and type of eruption. They include shield, composite, cinder cones, fissure eruptions, acid or dome volcanoes and calderas. Eruptions range from effusive to explosive.
- Volcanic processes and associated hazards include pyroclastic flows, lava flows, ash falls, lahars, jökulhlaups, landslides and toxic gases.
- The demographic, social and economic impacts of volcanic hazards depend on physical factors, such as the nature of the volcanic event, and human factors, such as population density, level of development, governance and mitigation strategies.
- Environmental impacts of explosive volcanic eruptions mainly affect weather and climate.
- Volcanic hazards have impacts at global, regional and local scales.

# ■ Earthquakes, processes, hazards and their impacts

## Earthquake characteristics, terminology and causes

Most earthquakes result from movement along fractures or **faults** in rocks. These faults usually occur in groups called a fault zone, which can vary in width from a metre to several kilometres.

Movement occurs along fault planes of all sizes as a result of stresses created by crustal movement. The stresses are not usually released gradually, but build up until they become so great that the rocks shift suddenly along the fault.

- As the fault moves, the shockwaves produced are felt as an earthquake by a process known as **elastic rebound**.
- The point of the break is called the **focus** (or hypocentre), which can be anything from a few kilometres to 700 km deep.
- If the stresses are released in small stages there may be a series of small earthquakes.
- Conversely, if the stresses build up without being released, there is the possibility of a 'Big One' — a major earthquake.

Often, as in the case of the Christchurch, New Zealand event, many of these faults are buried, so it is difficult to predict earthquakes when there is no knowledge of their existence.

During an earthquake, the extent of ground shaking is measured by motion seismometers activated by strong ground tremors, which record both horizontal and vertical ground accelerations caused by the shaking.

Analysis of data collected from the seismographs shows that earthquakes produce four main types of seismic wave, which are summarised in Table 12.

**Table 12** Types of seismic wave

| Seismic wave type | Characteristics |
|---|---|
| Primary (P) waves | P waves are vibrations caused by compression. They spread out from the earthquake fault at a rate of about $8\,km\,s^{-1}$ and travel through both solid rock (Earth's core) and liquids (oceans) |
| Secondary (S) waves | S waves move through the Earth's body at about half the speed of P waves. They vibrate at right angles to the direction of travel. S waves, which cannot travel through liquids, are responsible for a lot of earthquake damage |
| Rayleigh (R) waves | R waves are surface waves in which particles follow an elliptical path in the direction of propagation and partly in a vertical plane — like water moving with an ocean wave |
| Love (L) waves | L waves are similar to R waves but move faster and have vibration solely in the horizontal plain. They often generate the greatest damage, as unreinforced masonry buildings cannot cope with horizontal accelerations |

In an earthquake the ground surface may be displaced horizontally, vertically or obliquely, depending on wave activity and geological conditions. The overall severity of an earthquake is dependent on the amplitude and frequency of these wave motions. S and L waves are more destructive than P waves because they have a larger amplitude and force.

The recorded time intervals between the arrival of the waves at different seismogram stations are used to locate the **epicentre** (the point in the Earth's surface directly above the **focus** of an earthquake).

Three broad categories of earthquake focus, by **depth**, are recognised:

1 deep focus, 300–700 km

2 intermediate focus, 70–300 km

3 shallow focus, 0–70 km — these are the most common (around 75%) and cause the most damage

**Knowledge check 11**

Distinguish between the epicentre and focus of an earthquake.

# Earthquake processes and hazards

## Primary hazards

### Ground movement and ground shaking

**Surface seismic waves** represent the most severe hazard to humans and their activities, since buildings and other structures may collapse and kill or injure their occupants. Ground motion severs underground pipes and power lines, resulting in fires and explosions, especially from escaping gas (e.g. the 1907 San Francisco earthquake). Ruptured water pipes mean that often it is difficult to extinguish these fires.

Near the epicentre, ground motion is both severe and complex, as there is an interlocking pattern of both P and S waves and, theoretically, most damage should occur at the epicentre. Different surface materials respond in different ways to the surface waves, with unconsolidated sediments being most affected because they amplify the shaking. This leads to differential damage of buildings and infrastructure, based not only on distance from the epicentre but also on surface materials (local geological conditions). Steep topography, as in San Francisco, also amplifies 'waves'.

This differential damage was apparent in the Mercalli earthquake intensity levels for the 1989 Loma Prieta earthquake (MM 7.1). More than 98% of economic losses were a result of ground shaking, and 41 out of 67 deaths resulted when ground shaking caused the upper tier of the Nimitz freeway in Oakland, California to collapse because it was constructed on foundations of soft mud and bay-fill material.

The phrase 'buildings kill, not earthquakes' is meaningful when considering earthquakes with severe impacts. Building quality is key. Poorly built, unreinforced structures with heavy, tiled roofs are the most dangerous. In the 1988 Armenian earthquake (MM 6.9) 25,000 people were killed, 31,000 were injured and 500,000 were made homeless within a 50 km radius of the epicentre. Distance decay was clearly shown in that 88% of the older stone buildings were destroyed in Spitak, only 5 km from the epicentre, but only 38% in Leninakan, 35 km from the epicentre. However, in Leninakan, 95% of the more modern 9–12 storey Soviet-built pre-cast concrete frame buildings were destroyed (they had soft foundations and no earthquake proofing). In the 2008 Sichuan earthquake (MM 7.9), although the ground shaking formed a linear pattern extending along the Longmenshan fault, a large number of pre-cast concrete school buildings were completely destroyed, scattered over a wide area, with other buildings remaining comparatively unaffected. This hit the headlines because, as a result of China's one-child policy, many families lost their only child.

The **duration** of shaking is also important, longer periods of shaking causing more damage for events of the same magnitude.

> **Knowledge check 12**
>
> Explain how China's one-child policy exacerbated the tragedy of the 2008 Sichuan earthquake.

## Secondary hazards

### Liquefaction

Liquefaction is an important secondary hazard that is associated with loose sediments. This is the process by which water-saturated material can lose strength and behave as a fluid when subjected to strong ground shaking, which increases pore water pressures. Poorly compacted sands and silts situated at depths less than 10 m below the surface are most affected when saturated with water. After the earthquake the water sinks deeper into the ground and the surface firms.

In the earthquakes at Christchurch, New Zealand (2010), Mexico City (1985) and Valdez, Alaska (1964) liquefaction that caused buildings and infrastructure to collapse was a notable hazard, resulting in an almost random pattern of building destruction.

### Landslides, rock and snow avalanches

Severe ground shaking causes natural slopes to weaken and fail. The resulting landslides, rock and snow avalanches can make a major contribution to earthquake disasters, especially in mountain areas such as the Himalayas. These landslides hamper relief efforts, as in the 2005 Kashmir earthquake or the 2015 Nepal earthquake. It is estimated that landslides can double earthquake deaths, especially with high-magnitude earthquakes, because they can occur over a huge area.

Post-earthquake landslide risk varies with differences in topography, rainfall, soil and land use (whether forested or not). An example of an earthquake-generated rockslide occurred in Peru in 1970 as a massive rock avalanche broke away from the overhanging face of the Huascaran mountain. A turbulent flow of mud and boulders flowed down the Santa Valley, forming a wave 50 m high, travelling at an average speed of 70–100 m s$^{-1}$. The towns of Yungay and Ranrahirca were buried under debris 10 m deep, killing 18,000 people in four minutes. Flooding can occur, as it did in Sichuan when numerous landslides dammed temporary lakes, which subsequently burst through, causing flash flooding. The aftershocks of the 2017 earthquakes in central Italy generated avalanches after heavy snow.

### Tsunamis

Tsunamis are the most destructive secondary earthquake-related hazard. Most tsunamis are generated at subduction–converging (destructive) margins, with 90% of damaging tsunamis occurring in the Pacific Basin (hence the establishment of the Pacific Warning System, see page 96). Exceptions include the 2004 Indian Ocean tsunami in the Indian Ocean. The most active tsunami source area is around the Japan–Taiwan islands (over 25% of tsunamis).

Tsunamis occur if an earthquake rupture occurs under the ocean or in a coastal zone, if the focus is not deep within the Earth's crust, and if the magnitude of the earthquake (6+) is large enough to create significant vertical displacement. A tsunami is a series of ocean waves that 'spread out' from the earthquake focus, carrying large volumes of water, and debris too, once they reach land (Figure 16).

> **Knowledge check 13**
>
> Explain why only certain earthquakes cause tsunamis.

**1 Generation in deep ocean**

**2 Tsunami run-up**: nature of the waves depends on
(i) cause of the wave, e.g. earthquake or volcanic eruption
(ii) distance travelled from source
(iii) water depth over route
(iv) offshore topography and coastline shape

**3 Landfall**: impact depends on physical factors and land uses, population density and warning given. Waves radiate from the source in all directions

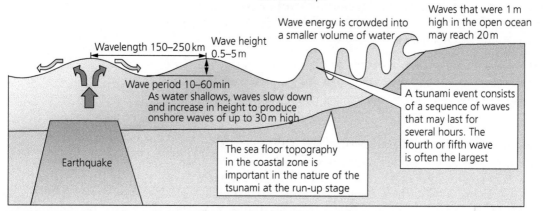

Figure 16 The formation and key features of a tsunami

The intensity (magnitude) of tsunamis can be measured by a descriptive, observational scale devised by Soloviev in 1978, which is based on the run-up height.

Over the last 100 years, more than 2000 tsunamis have killed over 500,000 people (over 50% of these in the mega-disaster of the 2004 Indian Ocean tsunami, the most deadly tsunami recorded).

A number of physical factors influence the degree of devastation, including wave energy, which is dependent on water depth, the process of shoaling, the shape of the coastline, the topography of the land and the presence or absence of natural defences, such as coral reefs or mangroves. Human factors include the population profile, the degree of coastal development, the cohesiveness of the society and people's experience of the tsunami hazard, as well as the presence or absence of warning systems and evacuation plans.

## Environmental, demographic, economic and social impacts of earthquakes on people and the built environment

There are a number of differences between the impacts of volcanoes and earthquakes, with their related tsunamis. In comparison with volcanoes, earthquakes can be much more deadly and their impacts more selective in terms of damage to buildings and deaths.

The **primary** effects of a major earthquake are the immediate consequences, such as damage to houses from shaking or fires, and instantaneous deaths of people hit by falling tiles and roofs. On the streets, cracks form across roads and bridges collapse; there is widespread destruction of gas mains and water pipes, and severe fractures or downwards concertina of badly built concrete high-rise blocks. Within a few minutes people are trapped and injured, with many dying quickly. Between 2000 and 2017, over 800,000 people died as a result of earthquakes, mainly due to collapsed structures.

Changes to the environment can result from landslides and from tsunamis associated with earthquakes. Land can also uplift or subside. In the 1926 Alaskan earthquake some areas were permanently raised by 9 m, while others subsided by 2.4 m, allowing the area to flood.

The **secondary** effects of an earthquake are those that manifest in the days, weeks and even months after an earthquake event. Air pollution might result from burning fires and combustion from leaking gas mains. Contamination from sewage is another serious secondary danger, causing diseases such as typhoid or cholera as a result of a shortage of clean water. After the 2010 Haiti earthquake 738,979 cases of cholera were reported, leading to 421,410 hospitalisations and nearly 10,000 deaths. As railroad and telephone links are cut and airports are damaged, the lack of supply lines for rescue and recovery is another secondary consequence (a major problem following both the Haiti and 2005 Kashmir earthquakes).

How people cope and how quickly they can get back on their feet largely depends on whether the earthquake occurs in a high-income country, where there are contingency plans for all stages of the hazard management cycle, underpinned by financial support mainly from within the country. In low-income countries there are fewer resources for rescue and recovery, and a reliance on international aid.

The secondary effects have both social and economic consequences. Many factories and offices are so damaged that work cannot resume for some considerable time, leading to losses in wages, production, future orders and exports. The community might also be threatened by hunger and disease and possibly by social disorder from looting etc. as people desperately seek to survive.

In rural areas, farmland and crops can be seriously affected if drainage or irrigation systems are disrupted and fields are covered in rubble. Where landslides have blocked roads, farmers are not able to get their products to market. On the other hand, in urban areas the effects can be extremely severe because of the high density of buildings and the high value of the infrastructure. If buildings are completely destroyed, leaving large areas of derelict land, uncollected refuse and decomposing organic material can result in infestations of rats and flies. The 2010 Tohoku earthquake had exceptionally severe secondary effects as a result of the 'triple whammy' of earthquake, tsunami and the resultant nuclear disaster.

Table 13 shows how the impact of an earthquake can vary between countries at different stages of development.

**Exam tip**

For A-level you must learn two contrasting examples of earthquakes to demonstrate the varied degree of risk and impacts of earthquake activity. For AS you need to learn one example.

Table 13 Comparison of the Nepal and Alaska earthquakes

| | Nepal (Gorkha) | Alaska |
|---|---|---|
| Date | 25 April 2015 | 30 November 2018 |
| Magnitude (Richter/Mercalli) | 7.8/VIII (severe) 38 aftershocks greater than magnitude 4.5 | 7.1/VIII (severe) 80 aftershocks greater than magnitude 5 |
| Plate boundary | Eurasian and Indo-Australian plates | Pacific and North American plates |
| Depth (km) | 15 | 47 |
| Deaths | 8964 (most villagers were working outside when the quake struck, reducing the death toll) | 0 |
| Injured | 23,447 | 0 |
| Homeless | 3.5 million | 110 people housed in hotels |
| Other impacts | 1/4 of the population affected Landslides — an avalanche on Mt Everest killed 21 824,000 homes destroyed Reduced harvest as affected farmers had little time to plant before the monsoon rains arrived Damage estimated at $5–10 billion Reconstruction cost $9 billion By 2017, only 5% of houses rebuilt | Some areas of liquefaction Thousands of homes had some damage Roads and highways damaged 46,000 left without electricity Damage estimated to be $76 million Tsunami warning issued then cancelled Trans-Alaska pipeline shut down for 7 hours as a precaution |

**Summary**

- Earthquake characteristics include focus, epicentre, depth and different types of waves (P, S, L and R waves).
- Earthquakes result from movement along fractures or faults in rocks and create different types of hazard, including ground shaking, liquefaction, landslides, avalanches and tsunamis.
- Environmental, demographic, economic and social impacts of earthquakes result from both primary and secondary causes, and have impacts at global, regional and local scales.

**Exam tip**

Always research up-to-date examples of earthquakes. Remember that the scale of the impacts may be adjusted as more information about recent events becomes available.

# Human factors affecting risk and vulnerability

## Disaster versus hazard

A disaster is the realisation of a hazard 'that causes a significant impact on a vulnerable population'. While the terms 'hazard' and 'disaster' are often used casually or synonymously, there is a major distinction between them, which is shown clearly by the Degg model (Figure 17).

**Knowledge check 14**

Explain the difference between a hazard event and a disaster.

**No hazard or disaster**

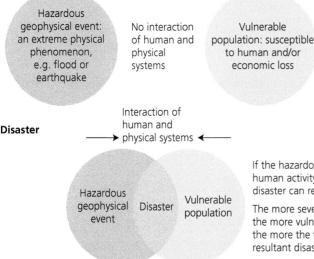

Hazardous geophysical event: an extreme physical phenomenon, e.g. flood or earthquake

No interaction of human and physical systems

Vulnerable population: susceptible to human and/or economic loss

Human activity and physical processes do not interact and there is no hazard or disaster

This would be the case with a volcanic eruption on a remote unpopulated island, or a landslide in an unsettled area

**Disaster**

Interaction of human and physical systems

Hazardous geophysical event — Disaster — Vulnerable population

If the hazardous geophysical processes and human activity become closer together, then a disaster can result

The more severe the geophysical event and/or the more vulnerable the human population, the more the two overlap and the larger the resultant disaster

**Figure 17** The Degg model

## Risk

**Risk** is defined as 'the probability of a hazard occurring and creating loss of lives and livelihoods'. It might be assumed that risk of exposure to tectonic hazards is involuntary, but in reality people consciously place themselves at risk for a variety of reasons, including:

- the unpredictability of hazards — areas may not have experienced a hazard in living memory
- the changing risk over time (e.g. a perceived extinct volcano)
- lack of alternative locations to live, especially for the poor
- an assessment that economic benefits outweigh the costs (e.g. for areas of rich volcanic soils or of great tourism potential)
- optimistic perceptions of hazard risks — 'it can all be solved by the technofix' or 'it will not happen to me'

**Exam tip**

Research White Island, New Zealand, and the eruption in 2019. It is a good example of people placing themselves at risk from tectonic activity.

The risk is altered by human conditions and actions, for example two similar-magnitude earthquakes (Loma Prieta, California and Bam, Iran) had very different consequences because the people in Bam (poorer in a developing country) were generally at much greater risk, i.e. they were more **vulnerable**.

# Vulnerability

**Vulnerability** implies a high risk of exposure to hazard, combined with an inability to cope. In human terms this is the degree of resistance offered by a social system to the impact of a hazardous event. This depends on the **resilience** of individuals and communities, the reliability of management systems and the quality of governance that has been put in place.

Certain conditions amplify vulnerability. These are outlined below.

## The risk equation

The risk equation measures the level of hazard risk for an area:

$$\text{risk} = \frac{\text{frequency and/or magnitude of hazard} \times \text{level of vulnerability}}{\text{capacity of population to cope (i.e. resilience level)}}$$

While the **intrinsic** physical properties of a hazard event profile can lay the foundations for the development of a disaster, it is the **extrinsic** areal or local factors that impact on the vulnerability of communities and societies and cause tectonic disasters. It is also to do with the actual communities and societies themselves.

The PAR model (Figure 18) helps to explain the variability in levels of vulnerability and resistance. It is this vulnerability (both human and economic), and not the tectonic environment, that helps to explain the differences in the severity of the social and economic impacts of physically similar hazard events (Table 14).

**Knowledge check 15**

Explain how unsustainable development can increase the risk equation value.

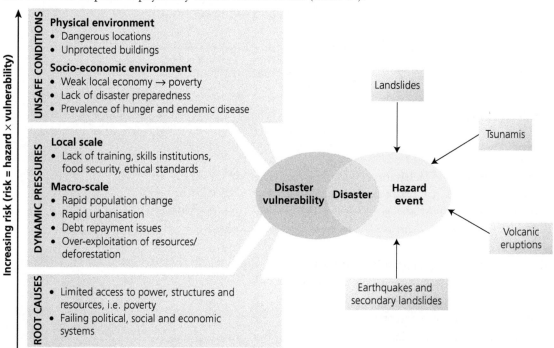

**Figure 18** The PAR (pressure and release) model

Table 14 Differences in the severity of social and economic impacts of physically similar hazard events

| Year | Event | Magnitude | Fatalities | Damage (US$ millions) |
|------|-------|-----------|------------|------------------------|
| 1992 | Erzincan, Turkey | 6.8 | 540 | 3,000 |
| 1999 | Izmit, Turkey | 7.4 | 17,225 | 12,000 |
| 1989 | Loma Prieta, USA | 7.1 | 68 | 10,000 |
| 1994 | Northridge, USA | 6.8 | 61 | 44,000 |

As the PAR model shows, certain drivers of disaster (root causes) result in pressures that create potentially unsafe conditions. The development paradigm argues that, at a macro-scale, the root causes of vulnerability lie in the contrasting economic and political systems of the developed/developing divide. The most vulnerable people are channelled into the most hazardous environments (the result of chronic malnutrition, disease, armed conflict, chaotic and ineffective governance, and lack of educational empowerment). Therefore, the risk equation value is being increased because levels of vulnerability are increasing and resilience is decreasing.

## Drivers of disaster and vulnerability

### Economic factors

Human vulnerability is closely associated with levels of absolute poverty and the economic gap between rich and poor (inequality). Disasters are exacerbated by poverty (Haiti, Kashmir etc.). The poorest least-developed countries (LDCs) lack money to invest in education, social services, basic infrastructure and technology, all of which help communities overcome disasters. Poor countries lack effective infrastructure. Economic growth increases economic assets and therefore raises risk unless managed effectively. However, developed countries can invest in technology for disaster reduction and production, and in aid after the hazard event.

### Technological factors

While community preparedness and education can prove vital in mitigating disasters, technological solutions can play a major role, especially in building design and prevention and protection, and also in the design of monitoring equipment (page 94). However, in wealthy countries the reliance on technology may be much greater, making them more vulnerable to problems if the hazard affects the technology systems.

### Social factors

The world population is growing, especially in developing nations where there are higher levels of urbanisation and many people live in dense concentrations in unsafe political settings. It is not only the density of a population but also the population profile (age, gender and levels of education) that are significant. An increasingly ageing population, as in China (Sichuan), increases vulnerability through problems with emergency evacuation and survival. Housing conditions and quality of building have a major impact on the scale of deaths and injuries. Essentially, disadvantaged people are more likely to die, and suffer injury and psychological trauma during the recovery and reconstruction phase because they live in poorer housing that is not earthquake proof.

In the 2008 Sichuan and 2005 Kashmir earthquakes, badly built schools led to disproportionate deaths among the young.

In some societies, women may be more vulnerable due to fewer financial resources and the role of protecting other family members. Also, a lack of education about hazard risks can increase vulnerability. Unaware of the warning signs, some people walked towards the sea to take a closer look at what was happening prior to the arrival of the 2004 Indian Ocean tsunami.

### Political factors

The lack of strong central government produces a weak organisational structure. Equally, a lack of financial institutions inhibits disaster mitigation and both emergency and post-disaster recovery. A good, strong central government leads to highly efficient rescue (e.g. Chinese earthquakes). The 2015 Nepal earthquake is an example of government building standards having been ignored, often with extra floors having been added against code. Just 12 fire engines, many of which were old and out of service, provided for the area. Haiti is a classic case of the cumulative impacts of poor-quality governance over many years.

### Geographical factors

Geographical factors can be highly significant and case-study-specific, such as location (rural, urban, coastal), degrees of isolation and time of day.

- Increasing urbanisation, with poorly sited squatter settlements, especially in megacities, creates high hazard risk and exposure. These huge cities are vulnerable to post-earthquake fires (e.g. Kobe).
- Destruction of rural environments can result in disasters among rural populations, with a loss of food supplies and livelihoods (e.g. the 2015 Nepal earthquake).
- Relief, rescue and recovery efforts are difficult in some areas (e.g. Kashmir, where isolation, the cold climate and a frontier position complicated relief and recovery).
- The geography may lead to multi-hazard hotspots, where the impacts of earthquakes, tsunamis or volcanic eruptions are amplified by the impacts of other hazards.
- Timing of the first earthquake and aftershocks have a major effect, especially on social impacts, such as deaths. The 2015 Nepal quake occurred when many were working outside, reducing the death toll from collapsed buildings. The Italian earthquake of 2016 occurred during the summer holiday season, when the population in the area was much higher.

> **Knowledge check 16**
>
> Why does a hazard profile, as shown in Figure 11 on pages 68 and 69, not show the impact a tectonic event may have?

## Summary

- Economic factors affecting risk and vulnerability include level of development, level of technology, inequality and poverty.
- Social factors include population density, population profile (age, gender), housing conditions, quality of building and levels of education.
- Political factors include the quality of governance and strength of central government.
- Geographical factors include rural/urban location, time of day and degree of isolation.

# Responses to tectonic hazards

**Monitoring**, **predicting** and **warning** of tectonic hazards (Figure 19) are examples of modifying people's **vulnerability** to the hazard — this also includes community preparedness and land-use planning.

Prediction buys time to:

- warn people to evacuate
- prepare for a hazard event
- manage impacts more effectively
- help insurance companies assess risk
- prioritise government spending
- help decision makers carry out cost–benefit calculations of, for example, building expensive hi-tech systems

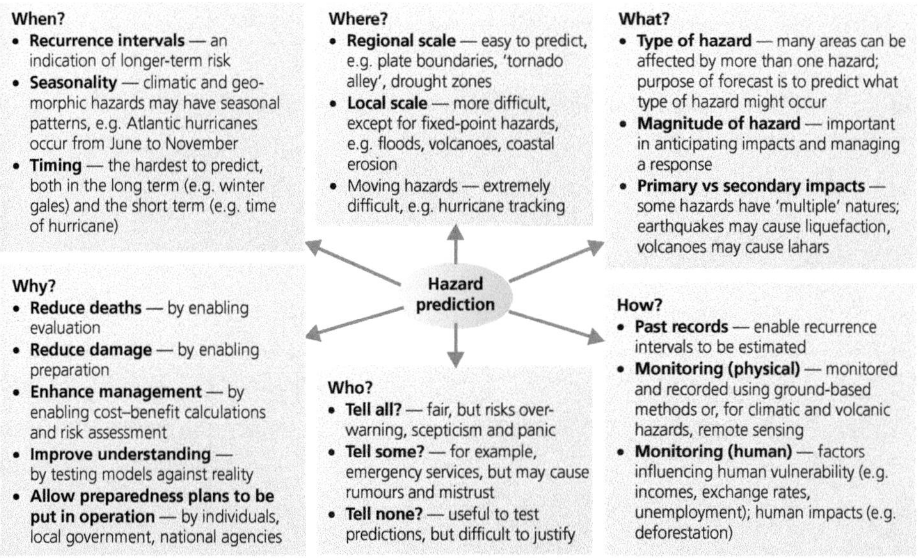

**When?**
- **Recurrence intervals** — an indication of longer-term risk
- **Seasonality** — climatic and geo-morphic hazards may have seasonal patterns, e.g. Atlantic hurricanes occur from June to November
- **Timing** — the hardest to predict, both in the long term (e.g. winter gales) and the short term (e.g. time of hurricane)

**Where?**
- **Regional scale** — easy to predict, e.g. plate boundaries, 'tornado alley', drought zones
- **Local scale** — more difficult, except for fixed-point hazards, e.g. floods, volcanoes, coastal erosion
- Moving hazards — extremely difficult, e.g. hurricane tracking

**What?**
- **Type of hazard** — many areas can be affected by more than one hazard; purpose of forecast is to predict what type of hazard might occur
- **Magnitude of hazard** — important in anticipating impacts and managing a response
- **Primary vs secondary impacts** — some hazards have 'multiple' natures; earthquakes may cause liquefaction, volcanoes may cause lahars

**Hazard prediction**

**Why?**
- **Reduce deaths** — by enabling evaluation
- **Reduce damage** — by enabling preparation
- **Enhance management** — by enabling cost–benefit calculations and risk assessment
- **Improve understanding** — by testing models against reality
- **Allow preparedness plans to be put in operation** — by individuals, local government, national agencies

**Who?**
- **Tell all?** — fair, but risks over-warning, scepticism and panic
- **Tell some?** — for example, emergency services, but may cause rumours and mistrust
- **Tell none?** — useful to test predictions, but difficult to justify

**How?**
- **Past records** — enable recurrence intervals to be estimated
- **Monitoring (physical)** — monitored and recorded using ground-based methods or, for climatic and volcanic hazards, remote sensing
- **Monitoring (human)** — factors influencing human vulnerability (e.g. incomes, exchange rates, unemployment); human impacts (e.g. deforestation)

**Figure 19** The importance of hazard prediction

## Monitoring, prediction and warning of earthquakes

The United States Geological Survey (USGS) states that an earthquake prediction should define the following:

- the date and time
- the location
- the magnitude

Predicting earthquakes would allow people to evacuate the danger area before the event, but unfortunately seismologists are unable to fulfil this.

On a global scale, the regions of risk can be identified. At a **regional** scale, previous magnitude and frequency data can be used to pinpoint areas of risk and predict the **probabilities** of an earthquake occurring, but not precisely when this might happen. As an earthquake results from the release of strain building up in the crustal rocks, the areas that have 'loaded' for some time are likely to move in the future.

Seismologists in California have produced earthquake probability maps for major fault lines, such as San Andreas, based on this '**gap theory**'.

At a local scale, attempts to predict earthquakes a few hours before the event are based on diaries of survivors (living histories) and the results from monitoring equipment. This includes changes in groundwater levels, release of radon gas or even (often used in China) unusual animal behaviour. These changes are thought to be due to ground dilation and rock cracking just before an earthquake. While the 1975 Heichang earthquake was successfully forecast 5½ hours before the event, allowing 90,000 people to be successfully evacuated, the Chinese failed to forecast the Great Tangshan earthquake of 1976 (an intra-plate earthquake), which was totally unexpected and resulted in a huge death toll.

It is possible to produce earthquake risk maps, which predict the likely impact of an earthquake on a particular designated area. These maps are based on known factors such as rock types, ground shaking, hill slopes, liquefaction danger and landslide potential. GIS is being used to develop this system further, especially by mapping concealed faults.

Countries such as Japan and Mexico have started using earthquake early warning systems (EEW), which send out warnings minutes or seconds before the event.

Figure 20 shows a range of possible monitoring methods that could be used to explore earthquake processes.

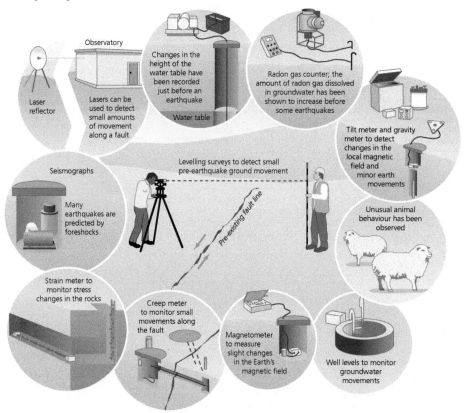

**Figure 20** Monitoring methods for predicting an earthquake along an active fault line

# Monitoring, prediction and warning of volcanic eruptions

Given adequate monitoring, warning of certain volcanic eruptive phases is possible, making adaptive responses such as education, community preparation and emergency evacuation procedures feasible.

Most volcanic eruptions are preceded by a variety of environmental changes that accompany the rise of magma to the surface (i.e. precursors to an eruption). Unfortunately, it is very difficult to predict the precise timing of the actual eruption reaching danger levels. Moreover, for highly explosive eruptions, many of the phenomena are not present, and the precise timing of these eruptions is highly unpredictable.

- **Earthquake activity** is common near volcanoes and, for predictive purposes, it is important to measure any increase in activity in relation to usual background levels. This requires an analysis of historic data plus supplementary field data from portable seismometers. Warning signs include a swarm of higher-frequency earthquakes, reflecting the fracture of local rocks as magmatic pressure increases. Audible rumblings can sometimes be heard using sound instruments.

- **Ground deformation** is sometimes the forerunner of an explosive eruption, although it can be difficult to measure for explosive subduction volcanoes. Tilt meters and other survey equipment are used to measure changes in slope. Electronic distance meters (EDM) can measure the distance between benchmarks placed on a volcano to pinpoint when the magma is rising and displacing the ground surface.

- **Global positioning systems** (GPS) rely on satellites that orbit the Earth twice a day and constantly feed back information that allows the ground base to provide profiles. GPS receivers in the volcano can detect the build-up of pressure from rising magma.

- **Thermal changes** occur as the magma rises to the surface and increases the surface temperature. Ground observations of hydrothermal phenomena, such as increased discharge from hot springs, increased steam from fumaroles, increases in temperature of crater lakes or hot springs, or wilting of vegetation on the volcanic slope, can be supplemented and confirmed by thermal imaging from satellites.

- **Geochemical changes** can be detected in the composition of gases issuing from volcanic vents (increasing $SO_2$ or $H_2S$ content). Direct field sampling of gases escaping from surface vents is the usual method, but remote sensing has been used too. $SO_2$ injected high into the atmosphere can be measured by onboard satellite equipment, and behaviour of volcanic plumes can be monitored by weather satellites.

- **Lahars** have been monitored for years by local people but more recently videocams allow automatic detection systems. Seismometers detect ground vibrations from an approaching lahar, so an emergency message can be transmitted downslope to population centres, enabling short-term warnings and emergency evacuation.

Although there is no fully reliable forecasting and warning system, some success has been achieved in limiting deaths. For example, Phivolcs is a successful scheme developed in the Philippines to monitor the most active volcanoes in the most densely populated areas.

Once a volcano has erupted, for example the one in Montserrat, a danger zone is then delineated. When clear warnings of new volcanic activity are received, people are evacuated from the danger zone. The community is prepared in advance on evacuation routes, and temporary food and shelter are supplied. However, the length of time available for evacuation is unpredictable, and sometimes there are false alarms. For example, in Montserrat, 5000 residents were evacuated three times between December 1995 and August 1996. In the Bay of Naples, home to Vesuvius, 700,000 people living in major cities such as Naples are at risk from its eruption. Volcanologists and civil defence officials have drawn up an emergency evacuation plan. The operation is huge and even involves moving people to safety by ship, with detailed routes carefully planned. Also, volcanic risk maps, using three grades of hazard risk, have to be created for a variety of hazards such as pyroclastic flows and ash falls. However, evacuation strategies, if managed effectively, can save thousands of lives.

# Prediction and warning of tsunamis

Modifying vulnerability is the main response to tsunamis. Scientists can predict a possible tsunami by monitoring earthquake activity, with the aim of issuing warnings to vulnerable populations who can evacuate the area. Tsunami forecasting is possible using tide gauges, detection buoys and pressure recorders located on the ocean bed. Warning systems are well established in the Pacific, although frequent false alarms can lead to threat denial and financial loss.

## Global-scale warning systems

In 1948 the Pacific Warning System for 24 Pacific Basin nations was established, with its centre near Honolulu in Hawaii. Seismic stations detect all the earthquakes and their events are interpreted to check for tsunami risk. The aim is to alert all areas at risk within 1 hour. The time it takes for a wave to travel across the Pacific allows ample time to warn shipping and evacuate low-lying coastal areas. As not all earthquakes result in tsunami, it is a difficult decision whether to issue a warning. If the earthquake is larger than 7.5 (MMS) all locations within 3 hours' 'travel time' of the tsunami waves are put on warning alert to evacuate the coast, with areas 3–6 hours away put on standby.

There was no Indian Ocean warning system in place for the 2004 tsunami, although one has subsequently been developed, based in Indonesia and India.

## Regional-scale warning systems

Regional-scale warning systems aim to respond to locally generated tsunamis with short warning times, as these pose a much greater threat. Ninety per cent of tsunamis occur within 400 km of the source area, so there may be less than 30 minutes between tsunami formation and landfall.

Japan has the most developed system — the target is to issue a warning within 20 minutes of the approach of a **tsunamigenic** earthquake within 600 km of the Japanese coastline. Such a warning was issued for the 2011 Tohoku earthquake, giving residents up to 10 minutes to evacuate. However, the height of the tsunami wall was insufficient for the 40 m-high waves and despite the warning, many people were unable to escape. Over 19,000 were killed — many at evacuation sites, some of which were washed away. The detector and computer system was set up in 1994 so that wave heights and arrival times could be rapidly transmitted. The system was updated in 2013 as a result of the 2011 event.

There are three main difficulties to overcome when issuing tsunami warnings. First, the tsunami may destroy power and communication lines. Second, as at Aceh, Indonesia in 2004, many events occur too quickly to issue a warning. Third, warnings must be supported by effective land-based evacuation routes and community education.

## Mitigation and adaptation of tectonic hazards

A useful framework for classifying responses to tectonic hazards, developed by K. Smith, divides them into three categories:

1 modify the event
2 modify the vulnerability
3 modify the loss

### Modify the event

While little can be done to control most volcanic hazards, some progress has been made in controlling lava flows. Seawater surges were successfully used to cool and solidify the lava flows during the 1973 Eldfell eruption in Iceland, and stop it advancing on the harbour of Vestmannaeyjar. Explosives were used with some success on Mt Etna in Sicily to create artificial barriers to divert lava flows away from villages in the 1983, 1991 and 2001 eruptions.

Some attempts have been made to modify the impact of the tsunami hazard by hazard-resistant design. Defensive engineering works provide some protection, and the trend now is for a combination of hard engineering, using onshore walls in high-value urban areas, and the redevelopment of natural protection provided by coral reefs and mangroves in rural areas.

Being able to control the physical variables of an earthquake, such as duration of shaking, is unlikely in the foreseeable future, although human-induced earthquakes, such as those caused by dam construction or fracking, could be prevented by restricting developments in areas subject to seismic hazards.

The main way of modifying an earthquake event is by **hazard-resistant building design** to develop **aseismic** (earthquake-resistant) buildings, as the collapse of buildings is responsible for the majority of deaths, injuries and economic losses. There is no clear relationship between building age and damage, although recent quakes have shown that specially designed, high-specification aseismic buildings in California and Japan do perform well even in high-magnitude earthquakes.

There are currently three main approaches to hazard-resistant building:

- Sizing the structure appropriately to be strong and ductile enough to survive shaking, with an acceptable level of damage.
- Equipping it with base isolation (e.g. construction on top of flexible pads).
- The techno fix, which involves expensive aseismic designs, such as structural vibration control technologies.

These are ideal for important public buildings and key services such as utilities (e.g. hospitals and power stations), but they are too expensive for homes. This is an important consideration because 70% of the world's 100 largest cities (12.5% of the world's population) are exposed to significant earthquake hazards once every 50 years.

Level of development is a key factor, as only economically developed nations can afford to **enforce** the strict seismic and building codes that can reduce death rates. In many developing or emerging nations there may be notional codes, but there is corruption and little money or political will to enforce them, hence the significant collapse of school buildings in the 2008 Sichuan earthquake. Recently, low-cost aseismic buildings suitable for rural and urban areas have been designed using cheap local materials, such as wood and wattle and daub, and avoiding materials such as concrete lintels and corrugated iron, which cause death and injury.

Two problems with this approach are, firstly, that, as in Christchurch, New Zealand, many older buildings need 'retrofitting' to bring them up to current higher standards of earthquake-resistant design and, secondly, damage is often from a variety of causes, not just shaking.

## Modify the vulnerability

In addition to prediction, warning and monitoring, other key parts of this strategy include land-use planning and zoning, community preparedness and education.

**Land-use planning and layout** is crucial in mitigating the severity of impacts of all three tectonic hazards.

- Tectonic hazard risk maps identify the most hazardous areas, which can be regulated by building codes. Lessons learned from **major earthquakes** are incorporated into planning new developments or rebuilding.
- Avoiding overly high-density urban squatter settlements and providing public open space creates safe areas away from fires and aftershock damage. Considerable thought also needs to be given to the siting of public buildings, which should be preferably scattered in low-risk areas to reduce the chances of the total collapse of services (this is part of Tokyo's planning).
- In areas of **volcanic hazard** risk maps are deployed (though few are available in developing countries). However, in areas such as Hawaii, lava flow hazards have been mapped and can be used as the basis for informed land-use planning, such as avoiding valleys where flows are concentrated.
- In **tsunami-prone** areas, rezoning of low-lying coastal land can be an excellent defence. For example, in Crescent City, California, following tsunami damage from the 1964 Alaskan earthquake, the waterfront has been turned into public parks, with businesses moved to higher ground back from the shore.

**Community preparedness and education** form the core strategy of any programme to modify the vulnerability of people to tectonic hazard. Many volcanic events are preceded by clear warnings of activity. Preparation of the community through education about what precursors to look for, how to evacuate an area and how to develop resistance is key.

Community preparedness for seismic hazards is centred on preparing the general public to cope, and the emergency services and government to manage before, during and after the event. Experience of how people behave in earthquakes has contributed to devising recommendations for appropriate action, and earthquake drills are now widely publicised.

In California, there is increased emphasis on using **smart technology** to prepare the emergency services.

> ### Knowledge check 17
>
> List the precursors communities should be made aware of.

# Modify the loss

Essentially this has two major facets: **aid** and **insurance**. **Insurance** is mainly available in economically richer nations. The vast majority of people at risk from tectonic hazards do not have access to affordable insurance. It is largely commercial and industrial property that is insured against tectonic risk and disaster damage.

Insurance is a key strategy for economically developed countries. However, while individuals realise that the benefits of purchasing an insurance policy are enormous and could outweigh the costs of damage, insurers who are wary of huge payouts following major hazard events assess the risk and charge accordingly. They also force householders to take preventative measures such as refitting their houses. Private properties become uninsurable in high-risk areas and for poorer people, leading to governments taking over provision in some instances.

Humanitarian concern for disaster victims results in **emergency aid** flowing in from governments, NGOs and private donations. Aid is used at all stages of the hazard management cycle (see below) for relief, rehabilitation and reconstruction.

**Exam tip**

Useful case studies for this section include Haiti 2010, Nepal 2015 and the successes and failures of international aid following the 2004 Indian Ocean tsunami.

# Short-term and long-term responses to the effects of earthquake and volcanic hazards

There are two useful frameworks we can use to look at responses over time.

1 The **Hazard Disaster Management Cycle** (Figure 21) identifies a number of phases in the management of a hazard from immediate response, through rehabilitation, to recovery and the development of resilience via mitigation strategies. Various versions of the cycle show how the strategies of modify the loss, modify the event and modify human vulnerability fit in the cycle. Today, technology is of increasing importance in the management of all stages of the cycle.

**Knowledge check 18**

Define the following terms: 'resilience', 'recovery', 'rehabilitation'.

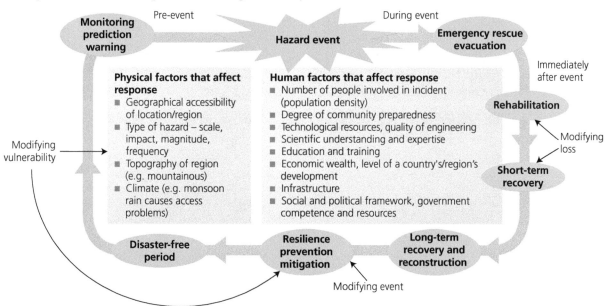

Figure 21 The Hazard Disaster Management Cycle

2 **Park's disaster–response curve** (Figure 22) allows modelling of the impact of a disaster from pre-disaster, through the impact, to post-disaster recovery, and shows the importance of various strategies over the life cycle of a hazard event.

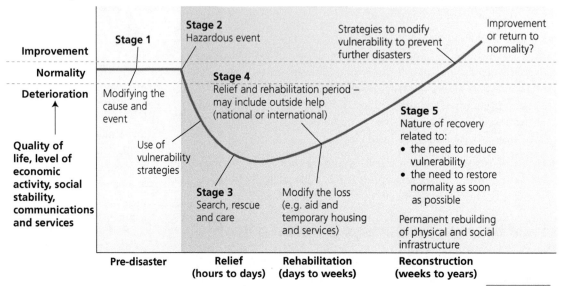

Figure 22 Park's disaster–response curve model

### Summary

- Monitoring, predicting and warning of tectonic hazards are ways of modifying the vulnerability of populations.
- Prediction can be effective for volcanic hazards, but is almost impossible for earthquakes.
- Short-term responses to the effects of earthquake and volcanic hazards include prediction and warning, and national and international emergency aid.
- Long-term responses include hazard-resistant building design, land-use planning, community preparedness, education and insurance.
- The Hazard Disaster Management Cycle shows how the choice of response depends on complex and interrelated physical and human factors.

**Exam tip**

Look at individual hazard events and devise 'tailor-made' diagrams for your chosen events to compare the progress and impact of both short- and long-term responses.

**Exam tip**

Use the internet to keep your case studies up to date — situations change continually. You could research the impacts of the 2004 Indian Ocean tsunami today. Use Google Earth to look at the rebuilding and see if you can still detect the damage. A comparison can be made with progress following the 2011 Tohoku earthquake and tsunami.

# Questions & Answers

## About this section

The questions that follow are typical of the style and structure that you can expect to see in the exam papers. Each question is followed by comments that give some guidance about question interpretation. Student responses are then provided, with further comments indicating the strengths and weaknesses of each answer and the number of marks that would be awarded.

When examiners mark your work, they use a grid that gives the maximum available mark for each assessment objective (AO). The mark scheme will have an indication of what should be included in the answer as well as marking guidance for the criteria required to reach the different mark bands.

You should always make use of examples where appropriate and reference data to support your answers. You can include sketch maps and diagrams where relevant. For AS exams the answers are written in the examination booklet, with the number of lines indicating the level of detail required. When writing in an answer booklet it is important to number your answers in the same way as the examination paper. If you use an extension page you must make a note such as 'continued on page…' at the end of the previous page. Remember to number the question on the extension page.

The formats of the different examination papers for this theme are given in the table on page 5.

# Coastal landscapes

## Question 1 (WJEC AS format)

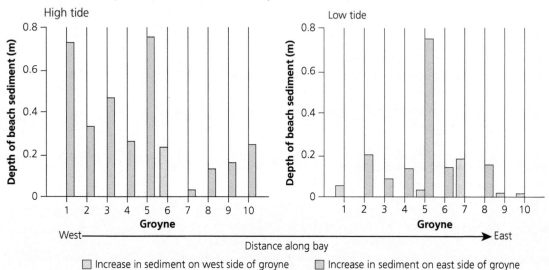

**Figure 1** Deposition of beach sediment along a coastal bay in summer

Students measured the difference in the amount of sediment that had been deposited alongside groynes in a bay. They recorded on which side of the groyne the greatest amount of deposition had occurred.

(a) (i) Use **Figure 1** to compare the differences in the deposition of beach sediment in different parts of the bay.

(5 marks)

> The command word 'compare' is targeting AO3 because it requires the interpretation and analysis of the information in Figure 1. You must provide a point-by-point identification of the similarities and differences shown by the data. You must compare and not simply describe each diagram. All 5 marks will be awarded for your use of skills to interpret and analyse the data (AO3).

### Student answer

Apart from groynes 7 and 8, the amount of deposition of sediment is greater in the high-tide zone compared with the low-tide zone. In some cases, such as groynes 1 and 5, there is 0.7 to 0.8 of a metre more sediment. In the high-tide zone all the build-up of sediment is on the eastern side of the groynes apart from groyne 6. This suggests that longshore drift is moving material from east to west. In the low-tide zone the number of groynes with most deposition on the eastern side is equal to the number where most is on the west, suggesting that longshore drift has less of an influence. In the high-tide zone the groynes in the western half of the bay have a much greater level of deposition than the eastern half. This is not so much the case at low tide, where the levels of deposition are similar, apart from the groynes at each end of the bay and groyne 5.

**5/5 marks awarded** This answer provides a thorough comparison of the two sets of data. It makes clear the differences and similarities in the amount of deposition as well as its location in relation to the side of the groyne and general location in the bay. Most comparisons are made using simple comparative terms such as greater, and there is correct use made of compass direction and one use is made of figures from the graph. This is more successful than writing a description of the high-tide zone followed by a description of low-tide, leaving the examiner to make the comparisons. At least five valid points have been made, so 5/5 marks for AO3 are awarded.

(ii) Suggest why a similar survey taken at the end of winter provided very different results. (3 marks)

In this question you are required to apply your knowledge and put forward plausible ideas about why there were differences. Therefore, all 3 marks will be awarded for AO2. The command word 'suggest' and the 3 marks available indicate that some elaboration is required, but not a lengthy explanation.

**Student answer**

During the winter there are usually more storms and windier weather than in summer. This can cause there to be more high-energy, destructive waves hitting the beach. These waves move sediment offshore, which may result in less material deposited and the removal of sediment in the high-tide zone, so that there is less of a difference between the high-tide and low-tide zones. Also, there may be a change in the prevailing wind direction during winter and as a result longshore drift may change direction or be reduced, resulting in less difference between the two sides of the groynes, especially in the high-tide zone.

**3/3 marks awarded** This answer shows a good understanding of how conditions influencing a beach may change with the seasons. This knowledge has been used to come up with some realistic explanations for changes that may occur. As at least three valid points have been made, 3/3 marks for AO2 are awarded.

(b) Examine the importance of aeolian processes in the formation of a coastal sand dune. (8 marks)

The term aeolian is used in the specification and so its meaning should be understood. This question requires you to demonstrate your knowledge and understanding of the formation of a sand dune (AO1). The use of the command word 'examine' indicates the need to consider the interrelationships involved in sand dune formation and evaluate the role played by aeolian processes (AO2). An answer should be supported with the use of a relevant example or examples. In this type of question, marks are not split evenly between the AOs. 5 marks will be awarded for the demonstration of your knowledge and understanding (AO1) and 3 marks for how you apply this knowledge to the question (AO2).

**Student answer**

A coastal sand dune is formed when winds blow onshore at the required velocity to be able to move sand particles inland by the process of creep and saltation. If there is an obstacle such as a rock or plant, this can reduce the speed of the wind so that the sand is no longer transported and begins to build up. Over time, more deposition may result in dunes and a ridge being formed, such as those at Studland in Dorset. It can be seen therefore that aeolian processes are important as without them the sand would not be transported inland.

However, there are a number of other conditions that need to be met in order for sand dune formation to occur. First, there must be an abundant supply of sand for the wind to move, deposited in the area by constructive waves. For example, the sand dunes along the Camargue coast of France are supplied with sediment carried by the River Rhone to the sea, which is then moved along the coast by currents.

In order for the wind to be able to move the sand easily, there needs to be a shallow beach gradient where a large area is uncovered at low tide, allowing the sand to dry out. There must also be an area where the blown sand can build up and where vegetation can become established to help stabilise the forming dunes.

Where all these conditions are present, it is possible for a sand dune system to form. It can therefore be assumed that aeolian processes are the most important factor required in sand dune formation. However, the formation is also dependent on a number of other factors without which the aeolian processes would have less impact.

**6/8 marks awarded** This is a competent attempt to establish the role of different factors in sand dune formation. It demonstrates a good level of knowledge by highlighting that factors other than aeolian are necessary, and attempts to demonstrate the interrelationship between the factors while also noting the premier role played by wind. There is some use of examples, although these are a little limited. Evidence from either Studland or the Camargue coast could have been used to show the necessity of other factors rather than just sediment supply. This has had an impact on the AO1 mark, limiting it to band 2. This answer is awarded 3/5 marks for AO1 and 3/3 marks for AO2.

## Question 2 (Eduqas A-level format)

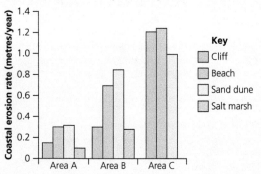

Figure 2 Erosion rates of different coastal landforms in three areas along the Canadian coast

(a) (i) Use **Figure 2** to analyse the variations in the rates of erosion at the locations on the Canadian coast. (5 marks)

The command word 'analyse' is targeting AO3, requiring you to interpret the data. You need to make clear the essential differences and should refer to the data as part of your answer. All 5 marks will be awarded for the interpretation and analysis of the diagram (AO3).

### Student answer

The rates of coastal erosion vary greatly between the three areas and also between the types of landform. The fastest rates of erosion occur in Area C, where the rates for cliffs and beaches are almost 1 metre a year faster. However, the rate for salt marsh is zero, although it may be because there is no salt marsh in that area. The slowest rates of erosion are in Area A, possibly because this area of coastline receives fewer high-energy waves than the others.

The unconsolidated material of beach and sand dunes erode much faster, apart from in Area C where cliffs also erode at a similar rate, possibly due to the rock type forming the cliffs. Salt marsh erodes at the slowest rate in all three areas, although it is only slightly slower than the rate of cliff erosion in Area B.

**3/5 marks awarded** This answer provides quite a well-developed analysis of the variations by looking at similarities and differences not only by area, but also by the type of coastal landform. This provides a better analysis than just describing the rates of each area in turn. It suggests some reasons for the differences but does not go into great detail, which is in line with the command word used. Some use is made of figures from the graph, but more could be made to fully show the scale of the variations. This answer achieves band 2 and is awarded 3/5 marks for AO3.

(ii) Suggest **one** way in which a change in sea level may impact on one of the landforms shown in **Figure 2**.

(2 marks)

The specification requires the study of only one type of sea level change and its impact on one landform. The command word 'suggest' requires the application of this geographical knowledge to come up with a plausible impact. The number of marks shows that there needs to be some elaboration of the impact, but does not ask for a lengthy amount of detail. As the question requires the application of knowledge, both marks will be awarded for AO2.

**Student answer**

If isostatic change results in a fall in sea level it can have a significant impact on the cliffs. The sea may no longer reach the foot of the cliffs, which may reduce the rate of erosion as they are no longer undercut. The cliffs may therefore become relict cliffs with features such as caves stranded above the high-tide level. Over time the face of the cliffs may become less steep due to the dominance of subaerial process acting on them.

**2/2 marks awarded** This is a good answer, suggesting a realistic change with adequate elaboration. This answer gains both AO2 marks, one for the suggestion and the second for the elaboration.

(b) Explain how coastal processes can have a positive impact on the growth of tourism.

(6 marks)

In this case the command word 'explain' is targeting AO1. To gain full marks you must demonstrate detailed and accurate knowledge of the topic. The specification makes clear the need to study the positive impacts of coastal processes on the growth of tourism. The answer should be supported with the use of appropriate and well-developed examples. All 6 marks will be awarded for AO1.

**Student answer**

A variety of coastal processes can help the tourism industry develop in an area. Erosional processes on certain coastlines produce dramatic and attractive scenery, such as cliffs, caves, arches and stacks, which may attract visitors to the area — for example, the Pembrokeshire Coast National Park. Sometimes distinct features may attract visitors, such as the 12 Apostles in Victoria, Australia. In such cases, the infrastructure for a tourism industry can develop because people are visiting the area.

Coastal processes creating bays and depositional processes resulting in the creation of beaches can also attract large numbers of visitors to coastal areas to help grow a tourist industry. This helps explain the growth of some of the traditional seaside resorts in England, such as Blackpool and Brighton, as well as the growth of tourism along the Spanish Costas.

Where processes allow the development of a specific ecosystem this may also become a tourist attraction. A good example of this is the Great Barrier Reef in Australia, which attracts almost 2 million people a year, putting the equivalent of over £2.5 billion into the Australian economy.

Coastal processes that may have happened a long time ago can also have a positive impact, encouraging the growth of a modern tourism industry. Changes in sea level resulting in the formation of a fjord has allowed Vancouver to develop as an important terminal for cruise ships, handling around 900,000 passengers a year, which puts $2 million into the economy.

As can be seen, a variety of coastal processes can have a positive impact on the growth of tourism by providing features that attract people to the area in the first place. This can then be developed further by the provision of economic activities to cater for the visitors. •

**4/6 marks awarded** This answer demonstrates quite a good level of knowledge about the topic without making the mistake of being too descriptive about the coastal processes. However, instead of using the general term erosional processes, more detail could have been provided. The answer does demonstrate how a variety of processes can impact on tourism and the findings are backed up with the use of examples, although these are not always developed enough for top marks. Consequently, it reaches band 2 and is awarded 4/6 marks for AO1.

## Question 3 (Eduqas A-level format)

Examine the success of **one** management strategy used to manage the impacts of coastal processes on human activity.

(15 marks)

> In this question you must demonstrate your knowledge and understanding (AO1) of a management strategy and apply that knowledge to evaluate the success of the strategy (AO2). You should develop a coherent line of reasoning, with a logical structure. In this type of question 10 marks will be awarded for AO1 and 5 marks for AO2.

### Student answer

One coastal management strategy is 'hold the line', where the present-day shoreline is protected in order to reduce or halt the rate of coastal erosion and thus limit its impact on human activity. a This strategy may involve the use of hard or soft engineering and is frequently used to reduce the impact of waves hitting the base of a cliff and so reducing undercutting and the possibility of cliff collapse. It is often used in areas where the coastline consists of weak rocks, which erode quickly.

a This answer sensibly chooses one of the five generic management strategies rather than writing about just one type of hard or soft engineering. This allows for greater use of examples and a more in-depth discussion about the level of success, which in turn gains both AO1 and AO2 marks.

One area with a hold-the-line strategy until 2105 is Walton-on-the-Naze in Essex. At the end of the 1970s around £1 million was spent building a sea wall, wooden groynes and cliff profiling in order to protect the residential area above the cliffs. Unprotected parts of the cliff have eroded at an average rate of 2 metres a year. However, where the management has been in place the shape of the cliffs has remained virtually unchanged, allowing the residential area to remain and actually grow in size. In this case, it can be seen that the management strategy has achieved its objectives and can be considered successful. The scheme has also greatly improved access to the beach as a result of steps and pathways built on the less-steep, reprofiled cliffs, which can also be considered a success. b

Other parts of the east coast of England have also adopted the strategy of hold the line. In 1990 the village of Mappleton in Yorkshire was under threat from coastal erosion. The coastline is made up of glacial till cliffs, which were eroding at 2 metres a year. In this case, £2 million was spent on a rock revetment and two rock groynes plus cliff stabilisation with landscaping and revegetation. As a result, a large beach has developed between the groynes. Combined with the revetment, waves reaching the base of the cliffs have been reduced, halting erosion and thus protecting the village. c In this case again it can be seen that hold the line can be a successful strategy.

However, there are some aspects of hold-the-line strategies that may not be seen as successful. One obvious failure was in 2014 when the sea wall at Dawlish in Devon collapsed, closing the main railway line and leaving homes under threat of flooding or destruction. d Repairs to the area cost millions of pounds and showed that such strategies are not fool-proof.

At Overstrand in Norfolk the combination of wooden revetment and groynes has slowed erosion, but the cliffs are still retreating, putting homes at risk, which shows that the hold-the-line policy has had limited success. e At Overstrand and at Mappleton the revetments have made access to the beach much more difficult. Although the main focus was to prevent erosion, this secondary impact may be seen by some as a failure. f

One frequent element of the hold-the-line strategy is the use of groynes to trap sediment moved by longshore drift. At Walton and Mappleton this has resulted in the amount of sediment reaching further along the coast being reduced, leading to increased rates of erosion. At Walton this resulted in £1.2 million having to be spent in 2011 on a policy of managed retreat next to the original scheme. In such cases it may be considered that the original strategy has had limited success. g

When the original strategy was adopted at Walton in the 1970s the cost of the scheme far outweighed the value of the property that was being protected at the time. Regarding cost–benefit analysis it might therefore be considered to be unsuccessful.

It can be seen that a hold-the-line management strategy can be frequently viewed as a success because it achieves its principle aim of preventing coastal erosion. However, the collapse of such schemes and some of the secondary impacts resulting from the scheme mean that describing such a strategy as an unqualified success must be questionable.

b c There is good demonstration of knowledge about the strategies in a variety of locations without the answer becoming too descriptive of the schemes. In each case the relevant information has been used to support the evaluation of the success of the strategy. There is a good demonstration of knowledge about the strategy in a variety of locations without the answer becoming too descriptive of the schemes, which counts towards AO1. In each case the relevant information has been used to support the evaluation of the success of the strategy, which gains credit for AO2.

d e The answer makes good use of examples to show where the policy has been less successful, which helps produce a balanced answer, demonstrating knowledge (AO1) and the ability to apply it (AO2).

f g This answer also sensibly considers the secondary impacts of the strategy, with examples, further demonstrating understanding for AO1. It helps develop the line of reasoning about the success of such a strategy, gaining credit for AO2.

**12/15 marks awarded** The answer has a logical structure and there is a coherent line of reasoning throughout, with a simple conclusion being reached. Detailed and accurate knowledge is shown (AO1), although perhaps some more detailed factual information about the secondary impacts could be added to help support the evaluation, which would enable the answer to gain top marks. The answer achieves band 3 and is awarded 8/10 marks for AO1 and 4/5 marks for AO2.

# ■Tectonic hazards

## Question 4 (WJEC AS format)

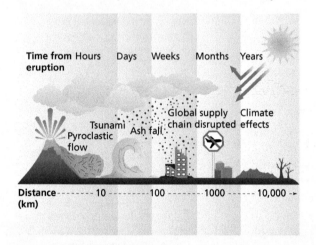

**Figure 1** The likely range of impacts of a large (VEI 6 or above) volcanic eruption

(a) (i) Use **Figure 1** to describe how the likely impacts resulting from a large volcanic eruption change with time and distance from the eruption. (5 marks)

> The AO3 command word 'describe' requires identification of the key changes in the impacts as shown in the diagram. You are using skills to interpret the diagram, rather than knowledge of actual examples, to write about the impacts. You must write clearly about both time and distance. All 5 marks will be awarded for the interpretation and analysis of the diagram (AO3).

**Student answer**

There is a variety of impacts that can occur from a short time span of hours to the long term, lasting years, and varying from the immediate area of the volcano to something that could be considered global.

The diagram shows that lava flows (although not labelled) and pyroclastic flows occur almost immediately following an eruption and impact an area up to 10 km from the source. It suggests that in the hours and days after the eruption a tsunami can affect an area up to 100 km from the volcano.

Another likely impact can be ash fall. The diagram shows that this can have an impact on built-up areas possibly up to almost 1000 km away and can last for days or for as long as almost months. It also shows that global supply chains can be affected for months, possibly as a result of planes being unable to fly through the ash clouds.

Finally, the diagram indicates that the impact which can occur furthest from the volcano and also the longest after the eruption is the effect on the climate. This can occur for years after the eruption and may impact thousands of kilometres away.

**5/5 marks awarded** This is a competent and descriptive account of the impacts, taking each one shown in turn and relating it to both time and distance. The student appears not to have assumed that each impact occurs only at the point where the feature is pictured or written, but that it extends from the volcano to the point where it is marked on the diagram. This answer does not make the error of including information that would form part of the answer to the next section of the question. At least five valid points have been made, so 5/5 marks for AO3 are awarded.

(ii) Evaluate the usefulness of **Figure 1** in describing the impacts of a volcanic eruption.  (9 marks)

This question requires the application of your knowledge about the impacts of volcanic eruptions to make a judgement about the suitability of the diagram. The answer should include both the strengths and weaknesses of the diagram. All 9 marks will be awarded for the application of knowledge and understanding to evaluate the diagram (A02).

**Student answer**

Figure 1 has a number of advantages and limitations in showing the impacts of a large volcanic eruption. The simplicity of the diagram is an advantage, which makes it easy to recognise that there are very different impacts and that they occur at different timescales and impact different areas. By using pictures, it allows a person to recognise the impact even if they do not know the correct term (e.g. pyroclastic flow). To some degree, it is possible to see that an eruption can have human, economic and environmental impacts by showing damaged buildings, disrupted supply chains and climatic effects.

However, the simplistic nature of the diagram can limit its usefulness. While showing some of the hazards that may occur, it does not show all of them. For example, it does not mention lahars or landslides, or make it clear that toxic gases could be emitted. It also does not differentiate between primary or secondary impacts. The diagram suggests that a tsunami is always a likely impact. While this is possible with a large eruption, this greatly depends on the location of the volcano and its proximity to the sea.

It would appear from the diagram that the impacts of a resulting tsunami would reach around 100 km from the volcano. However, the 2004 Indian Ocean tsunami (although caused by an earthquake) travelled thousands of kilometres from its source.

Likewise, the simplicity of the diagram means it could be interpreted either that climate effects do not occur for months or years and only occur thousands of kilometres away, or the effects cover thousands of kilometres and last for years. In reality, the effects could occur immediately and last for years, influencing the whole distance. While the diagram shows distance from the eruption, it might give a more realistic idea of an impact if it showed the aereal extent of an impact. This difference in interpretation could be applied to all the impacts shown.

Perhaps its greatest limitation that it does little to show the impact on people. A major eruption could result in large numbers of people killed, injured or left homeless, yet there is no indication of this in the diagram apart from some damage to buildings. It also gives no indication as to which of the features shown is likely to have the greatest impact on human life.

It can be seen that while its simplicity has some advantages, it also has many more disadvantages. Part of the problem is the title, because the diagram does not really show the impacts of an eruption but rather the potential hazards associated with it.

**6/9 marks awarded** This answer comprises quite a good evaluation of the diagram's usefulness, partially weighing up the pros and cons succinctly and providing a judgement, with a perceptive concluding sentence. The answer could benefit from the use of more supporting evidence rather than just the 2004 tsunami — for example, using details of an actual large eruption, such as Mt Pinatubo in 1991, to show how the scale of the impacts would compare to those shown in the diagram. The exam papers do mention the need to make the fullest possible use of examples to support answers. Mention could also be made of how the diagram is unable to show how the level of impact can be influenced by factors such as population density or level of economic development. The level of discussion and development of arguments about the diagram's usefulness allow the answer to attain band 2 with 6/9 marks awarded for A02.

**(b)** Describe how volcanoes form at converging (destructive) plate margins. (8 marks)

This question requires a detailed account and accurate demonstration of knowledge and understanding of how the volcanoes form, and is targeting AO1. It is not 1 mark for each correct fact. If a diagram is used it should have annotations that explain the formation or should be referred to in the written answer. All 8 marks will be awarded for the demonstration of knowledge (AO1).

**Student answer**

A converging (destructive) plate margin involves two plates moving together. Where an oceanic plate and continental plate converge, the oceanic crust, which is usually denser, is subducted down into the asthenosphere. The process of slab pull causes the plate to sink further, which pulls more of the plate into the subduction zone. As the plate sinks the heat from the mantle causes it to melt and a magma chamber forms in the area called the Benioff zone. If there are any lines of weakness in the continental crust above the magma chamber, pressure will cause the magma to rise and possibly reach the surface, forming volcanoes that can erupt violently. Examples of this include the volcanoes found in the Andes mountains.

If the converging (destructive) plate margin occurs where two oceanic crusts meet, a similar process occurs. However, the melting of the crust is often less deep below the surface, so magma has less distance to reach the ocean bed. Eventually, with more volcanic activity this can build up to form a volcanic island above the surface of the water.

**5/8 marks awarded** This answer demonstrates accurate knowledge of the processes resulting in the formation of volcanoes. In most cases it uses the correct technical terms. However, the use of examples is limited. Named examples of converging (destructive) plate margins and volcanic eruptions resulting from the subduction of a plate would be required to help achieve the top band of marks. Likewise, a named example of a volcanic island is also required. Further knowledge could be demonstrated by highlighting the lack of volcanic activity at collision margins where two continental crusts meet. However, you must take care to avoid the inclusion of too much irrelevant detail. As this answer demonstrates mostly accurate knowledge but with limited examples, it is awarded 5/8 marks for AO1.

# Question 5 (WJEC A-level format)

'It is easier to mitigate the impacts of volcanic activity than the impacts of earthquakes.' Discuss. (20 marks)

The question requires you to demonstrate your knowledge and understanding of the topic of mitigation. As well as factual knowledge, the 6 marks available for AO1 also include the use of exemplification and correct geographical terminology, supported by accurate spelling, punctuation and grammar. There are 13 marks available for AO2, where you must apply your knowledge to the development of the discussion. There is 1 mark for AO3 for an appropriately structured answer with a well-constructed argument and a conclusion.

**Student answer**

Both volcanoes and earthquakes can have serious impacts on both humans and the environment. Mitigate means to try to reduce the severity of the impacts, and humans have attempted to do this for tectonic hazards in a number of ways.

This short, simple introduction sensibly defines what is meant by mitigation, demonstrating understanding (AO1).

One method of mitigation is to try to modify the actual event so that the impact is lessened. For much volcanic activity this is not possible as the eruptions are too explosive. There have been examples where the flow of lava has been modified and so potentially reduces the impact. In 1973, 7 billion litres of seawater were sprayed on the lava flow from the eruption of Eldfell in Iceland, which helped solidify the lava before it reached the nearby harbour. On Mt Etna in Sicily they have used explosives for a number of eruptions to create barriers to divert the lava flow away from villages.

At present, the unpredictability and nature of earthquake activity has meant that humans have been unable to modify the actual event in any way. This shows that it can be easier to mitigate the impacts of a volcano compared with an earthquake, but only in limited aspects of the event. With a sudden explosive volcanic eruption the possibility of modifying the event may be just as impossible as modifying an earthquake.

Rather than just being descriptive about methods of mitigation, at the end of each point that is being made the answer refers back to the question to show how it supports or disagrees with the quote. As well as demonstrating understanding (AO1), it also demonstrates the application of knowledge and understanding (AO2).

Another way of mitigating impacts is to reduce the level of vulnerability of humans to the event. One method of achieving this is by monitoring and predicting when the event will occur, thus giving people the chance to take action to reduce the hazard impact. Volcanoes can frequently offer clues to the likelihood of an eruption, such as increasing seismic activity, changes in the emission of gases or a change in surface temperature. 🅐 This can allow warnings to be issued and people to be evacuated from the area, saving lives. This could be combined with education about preparing for an event.

It is still not possible to predict precisely where and when an earthquake will occur, and only warnings seconds in advance of an event have been attempted. It can be seen that this makes reducing the level of vulnerability much more difficult for earthquake activity and so, once again, mitigating the impacts is much more difficult. 🅒

Mitigation of the impacts of earthquakes can be achieved by the use of building codes that require the construction of earthquake-resistant buildings. These have been used in Japan and California and have helped to save lives when an earthquake has struck. 🅑 It is not so easy to do the same for volcanic activity. The variations in eruption type and scale means that buildings could be affected by the blast, pyroclastic flows, lava flows, lahars or large amounts of ash landing on them. In this case it would appear that it is not necessarily easier to mitigate the impacts of volcanic activity.

This section of the answer would benefit from better use of examples. At 🅐 there needs to be a named example to help support the point being made, while at 🅑 some more detail about how the buildings mitigate the effect would help. This would increase the AO1 mark. However, care must be taken to not make the answer too descriptive. At 🅒 and 🅓 the information has again been related to the quote to show if it supports or disagrees with it, gaining AO2 marks.

A further way to mitigate the impact is by hazard mapping, which can influence land-use planning. The areas of potential lahars or lava flows can highlight where not to build. Likewise, mapping areas of potential liquefaction, as happened in New Zealand, can prevent building in the area. In this case, providing the mapping is accurate and followed, there is little difference in the level of mitigation of the impacts of the events. 🅓

It is also possible to mitigate the impacts by modifying the loss caused by the event. With both a volcanic eruption and earthquake this may be by rescue and relief efforts, and could involve insurance or government and/or NGO aid. Providing the scale of the event is similar, it can be seen that the methods of mitigation in this case are similar, and it is not easier to lessen the impacts of volcanoes. ⓔ

Tectonic activity can also impact on the environment. With a large volcanic eruption, the aereal extent of its impact may be much greater, with large areas covered in ash, or particles or gas emissions circulating at high altitude and influencing the climate. On the other hand, the environmental impacts of an earthquake are usually much more localised unless it triggers the formation of a tsunami. Consequently, it can be easier to mitigate the impacts of an earthquake because you are dealing with a smaller area that has been affected. ⓙ

It can be seen that it is not always easier to mitigate the impacts of volcanic activity. However, the greater ability to predict a volcanic eruption means that it is easier to reduce the level of vulnerability through warnings and evacuations, and this can have the greatest effect on mitigating the impacts. Even so, the ease of this may be severely influenced by the nature and location of the event and the population's ability to respond.

At both ⓔ and ⓙ it would be better to give a named example to support the arguments being put forward to assist with gaining A01 marks.

**14/20 marks awarded** This answer clearly shows an understanding of what is meant by mitigating the impacts and it has demonstrated sound knowledge of how this can be achieved with both volcanic activity and earthquakes (A01). It successfully relates back to the question after each finding to show how it supports or disagrees with the statement (A02), which is better than writing descriptions of mitigation methods and leaving all the discussion to the last closing paragraph. The answer shows weakness in limited use of examples in providing details to support the arguments being made. This has affected the A01 mark by limiting it to band 2 (3/6). The answer has a good structure, putting forward different sides of the argument and reaching a satisfactory conclusion, thus gaining the A03 mark. This answer was awarded 3/6 marks for A01, 10/13 marks for A02 (band 3) and 1/1 mark for A03.

## Question 6 (Eduqas A-level format)

'Hazard profiling is the most important factor in reducing the impacts of tectonic hazards.' Discuss.

(38 marks)

With this type of question there are 14 A01 marks available for the demonstration of thorough and accurate knowledge of the subject and the use of well-developed examples. The 20 A02 marks are available for the application of the knowledge to answer the question. You must show a sophisticated application of knowledge to analyse and discuss the statement. You must develop a clear, coherent argument, logically structured and with a suitable conclusion, to gain the 4 marks for A03.

**Student answer**

All tectonic events can have a significant impact on an area, which may be social, economic or environmental. The scale of the impact can be due to the type of tectonic activity. One method of comparing tectonic events is to create hazard profiles. These show the main characteristics in the form of a diagram, such as the one below.

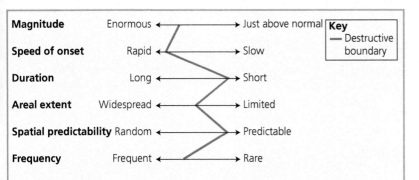

| | | | Key |
|---|---|---|---|
| **Magnitude** | Enormous ← | → Just above normal | — Destructive boundary |
| **Speed of onset** | Rapid ← | → Slow | |
| **Duration** | Long ← | → Short | |
| **Areal extent** | Widespread ← | → Limited | |
| **Spatial predictability** | Random ← | → Predictable | |
| **Frequency** | Frequent ← | → Rare | |

It is possible that being able to compare either different tectonic activities or different occurrences of the same type of event can help people understand the impacts and so can reduce them in the future.

This is a satisfactory introduction, as it notes that there can be different types of impact and explains what is meant by a hazard profile. This demonstrates knowledge, gaining AO1 marks. There is good use of a simple diagram to aid the explanation.

In December 2003 an earthquake hit Bam in Iran. It was a magnitude 6.8 quake, caused when two plates suddenly moved alongside each other. The earthquake resulted in over 26,000 people being killed with another 20,000 injured. More than 75,000 people were made homeless as over 80% of all the buildings in the area were destroyed. Services like electricity and water supply were destroyed, and teachers and doctors were killed. It was estimated that the economic cost of the event was almost $2 billion.

Another earthquake with a similar hazard profile occurred on 22 February 2011 when a magnitude 6.2 quake hit Christchurch in New Zealand. This may have been an aftershock from a bigger quake, which happened in New Zealand the previous year. The earthquake killed 185 people and 220 had major injuries and almost 7000 had minor ones. While few people were left homeless, over 10,000 buildings had to be demolished and thousands were damaged. The water supply and sewerage systems were destroyed. It was estimated that one-fifth of the population left the area permanently, and the economic cost of the event was $11 billion.

This is good, factual detail about two named examples (AO1). However, the information is descriptive and is not related to the quote, and so adds little to the discussion, which limits the AO2 mark.

Although the two earthquakes were similar in size and profile, it can be seen that they had very different impacts, so other factors must be important. The impact of a tectonic event can be influenced by many things. Economic factors can be important. In a poor country the people may be more vulnerable as buildings are poorer in quality and so more likely to collapse. They may not be able to evacuate if warnings are given because they cannot afford to survive elsewhere. A poor country [a] may not be able to afford warning systems or have a developed education system where the population is taught about the risks and how to respond. In Japan, pupils are taught how to minimise the risk from an earthquake. Poor countries are unlikely

A valid point is being made but it is not related to the concept of hazard profiling, so does not add to the discussion (AO2). A named example of a relevant poor country, for example at [a], would also demonstrate a greater level of knowledge (AO1).

to be able to deal with the effects and damage caused by an event, which means the impacts may last longer. In richer countries there are probably more things to be damaged because more technology is used and people own more, which means the economic cost of the damage may be greater.

Impacts of a tectonic event can also be influenced by social factors. In an urban area with a high population density, there is more potential for people to be killed or injured than if an event hit a sparsely populated rural area. Elderly people, the young and, in some societies, women may be more vulnerable due to factors such as the lack of physical strength or limited education about how to reduce the risks. Where the population is poor, they will have fewer resources to be able to deal with the damage and economic loss. Sometimes the impact of a tectonic event can be influenced by a person's perception of it. Some may adopt an 'it won't happen here' approach and so do little to prepare. In some cases, a person may ignore warnings because there have been so many false alarms in the past.

The quality of governance can affect the level of impact an event has. For example, where there is a lack of planning for the event or failure to enforce building regulations, the impact may be increased.

Geographical factors can have an effect. If an event occurs in a remote area the death toll may be lower due to a smaller population. However, it may be much harder to get aid to the area after the event. Even the time of day can have an influence. An earthquake at night may kill more because people are indoors when buildings collapse.

It can be seen that a number of factors can play a part in influencing the impact of any tectonic hazard. The type and size of the tectonic event can play a major role in the impacts, so understanding the profile of the hazard can help to reduce the impact in future. 🔟 However, it is also important to understand the role of other factors — economic, social, political and geographic — as these can all influence the impact.

There is plenty of factual detail above about the factors relating to the impact of a tectonic hazard (A01). However, it is not related back to hazard profiles and used to create a debate on whether the quote is supported. Again, this will impact on the A02 mark.

**19/38 marks** This answer demonstrates secure factual knowledge about hazard profiles and the factors influencing the impact of tectonic hazards. Where used, the examples are appropriate and the details accurate. It shows a sound understanding of what is meant by hazard profiling, although using the actual hazard profile of one of the examples mentioned could have been advantageous in helping to answer the question. This demonstration of knowledge allows the answer to reach the top of band 3 for A01. However, the weakness in this answer is in the application of the demonstrated knowledge. Rather than constructing a written debate and putting forward different arguments supported by examples, it contains two descriptive examples of hazard events and then descriptive, factual detail about the factors that affect the risk and vulnerability, and therefore the impact of a tectonic hazard. At the end of each point being made there is no link back to the quote in the question to show how it supports or disagrees with the statement. This reduces the A02 (8/20) and A03 (2/4) marks. While there is a conclusion that touches on the fact that hazard profiling can reduce the impact of future events 🔟, it does not conclude on its level of importance compared with the other factors, which is what the question required. This limits the answer to the middle of band 2 for A02 and to band 2 for A03. The answer is awarded 9/14 marks for A01 (band 3), 8/20 marks for A02 and 2/4 marks for A03.

## Coastal landscapes

1 Energy from the wind, waves and tides, and sediment from weathering and erosion processes.

2 Energy and sediment can move from the system to the surrounding environment.

3 Beach nourishment or 'feeding' involves humans adding pebbles and sand to a beach to replace those lost to longshore and offshore movement, so balancing the budget deficit.

4 Equilibrium is upset by changes in energy conditions, for example, during a storm (physical factor) or by rising sediment impact from human actions, both of which can lead to rapid changes over short periods of time.

5 Threshold is the critical water velocity at which a particular size of load is entrained or deposited. Fine material is very cohesive and therefore requires a higher threshold of current velocity than coarser sand.

6 These events are single occurrence events happening over a short period of time.

7 Wavelength ($\lambda$) is the horizontal distance between two crests or troughs. Wave amplitude is the height of the crest above stationary water — not the same as wave height, which is the vertical distance between a crest and a trough. Wave frequency is the number of crests or troughs passing a stationary point over a time period, such as a minute.

8 It cuts down the rate of erosion and therefore cliff recession rates as the waves do not break but are reflected back from the cliff, usually in deep water.

9 A rotational slide is a downslope movement of material that occurs along a distinctive curved slip surface. A slump is the result of the rotational slide, when a portion of the steep slope moves forward for only a short distance.

10 The platforms are formed by erosion and salt weathering during tidal exposure as well as by wave quarrying and abrasion, so the more general term is more accurate.

11 Sorting by waves and tides is illustrated by the Hjulström curve, whereby the deposits are entrained and deposited according to the velocity and size of the pebbles. An example is storm beaches, where only large boulders are hurled up the beach beyond high-tide level.

12 90% of sediment input comes from rivers. A dam can trap this sediment, reducing the amount available to be deposited in landforms.

13 This is a direct result of the size of the sand grains, which can easily be bounced along by the wind.

14 The loss of vegetation results in conditions making it more difficult for vegetation to become established, thus amplifying the change.

15 Physical factors: sediment supply can enhance or diminish the available silt; changes in river currents and volume can affect erosion; storms can erode the marsh; changes in tidal currents can increase erosion and alter species; changes in wave direction, nature and size can affect marsh stability; climate affects species types, growth rates and sea levels; sea level rises can upset equilibrium and destroy the marshes. Human factors: commercial, industrial and recreational activities can damage the marsh.

16 This 1-in-200-year event was the result of extreme physical factors (storm surge 1.3 m, extremely high spring tides and very strong 130 km h$^{-1}$ onshore winds) combined with unfavourable human factors (poor-quality coastal defences [railway owned] and also widespread building of holiday bungalows and mobile homes on the flat coastal plain).

17 Eustatic change is when the sea level changes worldwide because of the volume of water in the oceans, for example, because of climate change. Isostatic sea level changes result from an increase or decrease in the height of the land, as a result of ice ages, sediment deposition or tectonic activity, and occur locally.

18 A ria is a submerged non-glaciated valley and a fjord is a submerged glaciated valley.

19 Rias provide deep water ports for industry and allow vessels to travel inland. Tidal mudflats can be drained for agriculture. Marine terraces can provide flat areas for road and rail communications. Attractive scenery can encourage residential land use.

20 Statistical techniques, through a process called frequency analysis, are used to estimate the probability of the recurrence of a given precipitation or flooding event using data analysis. So this event has a 1% chance of occurrence in any given year.

21 Where high-value and high-risk installations such as nuclear power stations are under threat from rapid erosion.

22 Movement of sediment is self-contained within each cell. Impacting on the input or movement of sediment may influence the processes and landforms elsewhere within the cell.

23 Many aspects within a coastal zone are considered. Environmental, economic, social, cultural and recreational considerations are taken into account, while aiming for sustainability.

24 It avoids hard engineering solutions and retains the natural processes of estuary zones. It also provides conservation potential as the new habitats will develop into high-quality, ecofriendly environments. It avoids coastal squeeze. It is a cheaper option, so there are opportunities to offset costs with income from tourism. Benefits should be evaluated against potential problems.

25 Sand dunes form a very fragile environment, and quickly adjust to sediment inputs and wind patterns.

Where there is bare ground the unconsolidated sediments are susceptible to wind erosion. This is especially true on embryo dunes and foredunes, where pioneer vegetation leaves much bare ground. Storm events can quickly erode the embryo dunes. Human activity can reduce vegetation cover, which again increases the possibility of erosion.

## Tectonic hazards

1 The mantle
2 **a** The most violent earthquakes are found at converging converging (destructive) margins and conservative (transform) margins.
  **b** The most explosive volcanoes are found at converging (destructive) margins or are supervolcanoes associated with continental hotspots.
3 A fault scarp is the initial slope formed by faulting, which is eroded to form a fault-line scarp over time.
4 A period of fold mountain building.
5 The difference is one of location. When molten rock is still located within the Earth it is known as magma. When molten rock reaches the surface and is extruded, it is known as lava.
6 A smaller earthquake or tremor that follows a major earthquake. In Christchurch 2011, the second earthquake was thought to be an aftershock. Like many aftershocks it was significant because it occurred in the city centre, so causing high damage.
7 The theory suggests that earthquakes can be predicted according to the relative size and frequency of earthquakes in a given area. An active fault zone should have a similar amount of movement along its length. Therefore, an area that experiences many small earthquakes is unlikely to experience a large one. However, if an area of a fault zone has not experienced an earthquake for a long time, it is more likely to experience a larger quake in the future.
8 Ash fall is volcanic ash that has been emitted in an eruption, falling back to Earth. A pyroclastic flow is where volcanic ash and fragments mix with gas and travel down the sides of a volcano at high speed.
9 Using Tohoku 2011 as an example, the primary hazard was the earthquake, the secondary hazard was the tsunami and the tertiary hazard was the nuclear power station disaster caused by the flooding from the tsunami.
10 An active volcano is one that has had at least one eruption during the past 10,000 years. A dormant volcano is one that is not erupting nor is expected to, but could erupt again in the future as it has in historic times. An extinct volcano has not had an eruption for at least 10,000 years and is not expected to erupt again in a comparable timescale in the future.

11 The difference lies in location. The focus is the exact point inside the crust of the Earth where the quake begins. The epicentre is the point on the Earth's surface directly above the focus.
12 The earthquake killed many students in schools because of timing and poor building quality. For many families their only child was killed.
13 Certain conditions are required in combination: fault and uplift in a coastal zone or nearby ocean, high magnitude (MM 6+ earthquake), vertical displacement, and shallow focus of earthquake.
14 A hazard event may result in damage, injury and death. When these impacts seriously disrupt the functioning of a community it is a disaster.
15 Increased building and higher densities of population in high-risk cities, with these areas often in shanty towns; deforestation increasing landslide risks after earthquakes; removal of protective vegetation (mangroves and corals), which protect coastlines from tsunamis.
16 The profile shows only the physical features of the event. It does not show human factors, such as population density, level of development and vulnerability, which can all influence the type of impact.
17 Possible precursors include any changes from the norm, such as small-scale, frequent earthquakes, bulging on the side of the volcano, changes in gas emissions, ground deformation, and changes in the water temperature of streams. Unusual behaviour in animals can also be considered.
18 Resilience refers to toughness, coping capacity and quick recovery time. Recovery involves the stages in getting over a disaster — immediate, short term and long term. Rehabilitation refers to the ability of communities to overcome the psychological upset caused by a disaster.

26 It is difficult to reconcile the conflicts between environmental, economic and social sustainability and the views of multiple stakeholders, who place increasing pressures on the coast, in contrasting locations and with different needs.